SO-BJT-127

BUILD YOUR OWN METAL WORKING SHOP FROM SCRAP

THE METAL SHAPER

Written & Illustrated

By

DAVID J. GINGERY

Printed in U. S. A.

FIRST EDITION

FIRST PRINTING 1981
SECOND PRINTING 1982

Copyright © 1981 DAVID J. GINGERY
all rights reserved

LIBRARY OF CONGRESS CATALOG
CARD NUMBER 80-66142

INTERNATIONAL STANDARD
BOOK NUMBER 0-9604330-2-3

DAVID J. GINGERY
" METAL SHOP FROM SCRAP "
2045 BOONVILLE
SPRINGFIELD, MO. 65803

CONTENTS

        THE METHODS AND MATERIALS THAT ARE SUGGESTED IN THIS
MANUAL WERE DEVELOPED BY A NON-PROFESSIONAL.   THE AUTHOR
IS NOT AN ENGINEER OR SCIENTIST.   NO LIABILITY IS ASSUMED
FOR INJURY TO PERSONS OR PROPERTY THAT MAY RESULT FROM THE
USE OF THIS INFORMATION.

FOREWORD

This simple series, which began as a hobby, has become a full time occupation. The response from shop hands all over the U. S. and Canada has been gratifying, and I'm very pleased to find so broad an interest in this type of activity.

In a sense, this group of projects is very much like retracing the steps of the pioneers of the machine tool industry. To begin with so little, and to acquire a group of very practical and durable machines is a little bit like a dream.

Even so, that's how it happened in history, and that's how it's happening in this series. The 3/8" electric hand drill and some energetic hand work built the lathe of home made castings, and the lathe has done the lions share of the work on the metal shaper. The future looks bright, because we have the shaper to help the lathe with the rest of the projects. I can really see no end of possibility, and the dream is a reality.

The metal shaper is my favorite project. Not only because it is so valuable and useful as a shop machine, but also because it provides some excellent problems in casting and machining on the lathe. The development of skill and knowledge that comes from this project will serve you well in your future shop activity.

Like the metal lathe, the end result is larger and more elaborate than I first thought possible. Such features as the automatic cross feed, adjustable stroke, rotating head, and the overall soundness of the machine are a real bonus. I had a concept of a much simpler project, and these possibilities did not present themselves until I began to design the machine. It is the home made castings that make so much possible. There is hardly a limit to what you can do with a compact home foundry.

Don't be intimidated by the machines appearance, it is not so complicated as it looks. While it involves a lot of work, the rewards are great. It will serve you for years with its ability to form complex metal parts.

INTRODUCTION

WHAT IS IT?

Having been considered obsolete in modern metal shop practice for many years, the metal shaper has very nearly disappeared from the shop scene. I don't know if any American manufacturer offers a small bench shaper at this time.

It was invented early in the industrial revolution, because a machine that could rapidly produce a true flat surface was so desparately needed in the manufacture of steam engines and machinery. Like the lathe, and all other machines, it evolved into a complex and elaborate machine that could do near anything.

The engraving below is a 24" floor model, made by The Hendey Machine Co. of Torrington, Conneticut. It was sold by the Hill & Clarke Machinery Co. of Boston. It was made of cast iron, weighed 2,200 pounds, and the price with the vise was $450.00 in 1888.

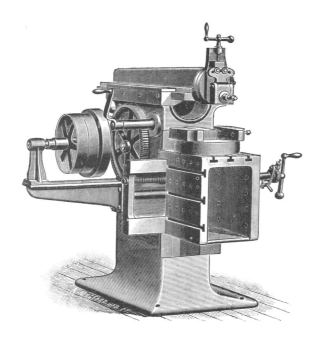

The type we are building is a column crank shaper. That means that the ram and the head are supported on a column, and it is driven by a crank. The work table travels on a horizontal slide.

On another type, called a traverse head shaper, the work table is stationary, and the reciprocating ram is carried on a horizontal slide.

Some designs used a rack and pinion to drive the ram, just like a planer, and others used a hydraulic cylinder.

It is essentially a planing machine, using a single point tool to make a straight line cut.

WHAT IS IT GOOD FOR?

As mentioned earlier, the shaper is considered obsolete. While it may not be the appropriate machine for the mass production of identical parts, it certainly has a lot to recommend it for the small shop.

It uses ordinary lathe tool bits, which cost from $2.00 to $3.00 each. The bits can be ground to any desired shape, and both ends can be used. Compare this to the cost of any single purpose milling cutter, and you will see that the shaper is a machine that you can afford to own and operate.

It is simple to set up and operate, and there are many jobs it can do better than other machines. Here are some of the jobs it can do: Flat surfaces, concave surfaces, convex surfaces, odd shaped profiles, keyways, splines, gear teeth, fluting for taps and reamers, slotting, and angular cutting such as used on a dovetail slide.

In my opinion, it is the most useful addition for a shop that does small one of a kind jobs like we do in the home shop.

HOW DOES IT WORK?

The tool head is fitted to the ram, which slides in very closely fitted ways in the column.

The ram is driven in a straight line by the crank, which is adjustable to vary the length of the stroke.

The tool is mounted on a clapper block, which allows it to lift on the return stroke. The cut is made on the forward stroke.

The work is mounted on a table which slides on horizontal ways.

The tool head is mounted on a pivot so that it can be set for angular cutting.

The crank pin is locked at any position in the slot in the crank plate, to vary the length of the stroke. The sliding block travels up and down in the yoke, causing it to move forward and back to drive the ram.

The opposite end of the crank shaft has another crank which operates the automatic cross feed ratchet. Its stroke is also adjustable to vary the amount of feed with each return stroke.

The table can be raised or lowered on rigid ways, and down feed is by a screw on the tool head, just like the compound feed on the lathe.

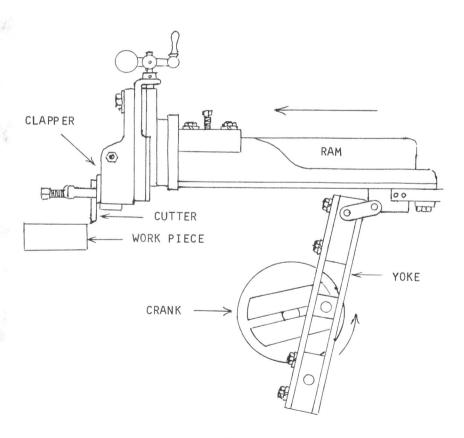

CLAPPER

RAM

CUTTER

WORK PIECE

YOKE

CRANK

SUPPLEMENTARY READING

A very thorough little manual, entitled " How To Run A Shaper " is published by South Bend Lathe. Like their manual on the metal lathe, it's a compact manual full of clear and useful information. If you have never run a shaper, it will serve as an excellent operators manual for the one you are about to build.   It's available from Lindsay Publications, PO Box 12, Bradley, Ill. 60915.

If you don't have a copy of South Bends " How To Run A Lathe ", you should order one along with the shaper manual. They are both excellent manuals, and worth much more than the small price.

Any of C. W. Ammens manuals on metal casting will be a valuable aid in understanding the foundry tasks in the project.   They are available from Lindsay, or from C. W. Ammen, PO Box 288, Manitou Springs, Colorado, 80829.

I'm not trying to sell you books, these are manuals that I used constantly while I was putting these projects together, and they will be a great help to you if you are new to metal working.

Your most valuable shop tool is knowledge, you simply can't know too much.

6 Inch Stroke Hand and Power Shaper.

The engraving is of a traverse head shaper made by the Boynton & Plummer Co. of Worcester, Mass. Made of cast iron, it weighed 350 pounds. It sold for $110.00 in 1888

CHAPTER I

PREPARING TO BUILD

NO SPECIAL EQUIPMENT NEEDED

When viewed as a finished product, the shaper may appear to be beyond the scope of a limited home shop operation. It is really a series of basic operations, and you will find no need for exotic equipment.

Only the home built lathe, from book 2 in this series, and a 3/8" electric hand drill were used in the construction of my shaper.

The design was limited by the capacity of the charcoal foundry, using a one quart pot, and the 7" X 12" capacity of the home built lathe.

FITTING OUT THE LATHE

I've often heard the comment that a metal lathe with no more than a face plate and a pair of centers is nothing but an ornament for the work bench. If you believe this, I am sure that you will change your mind before this project is finished.

Many a man has bought a metal lathe with all the money he could spare, only to find that the tools and accessories were going to cost more than the machine itself.

It would be well to realize that most tools and accessories are merely commercial versions of devices that were originally made by hand with limited equipment.

With the simple devices described in this section, you can easily perform all of the machining operations for the project on your home made lathe, or any other simple metal lathe that has no more than a face plate and centers.

MACHINING FLAT SURFACES

There are a number of the parts that must have a flat surface machined true and parallel. Much of the work can be done by simply bolting or clamping it to the face plate and facing it off with the cross feed. Some of the boring jobs can be done in the same way, but you will reach a point where you think you need a four jaw chuck.

9

MAKE A FACE PLATE CLAMP

These simple angle plate clamps will handle any of the jobs in the project that would require a four jaw chuck. Once you've made and used them, you'll be glad you didn't blow your bait and beer money on a chuck. We'll get into the construction of chucks in book 6, so save your money for more important things.

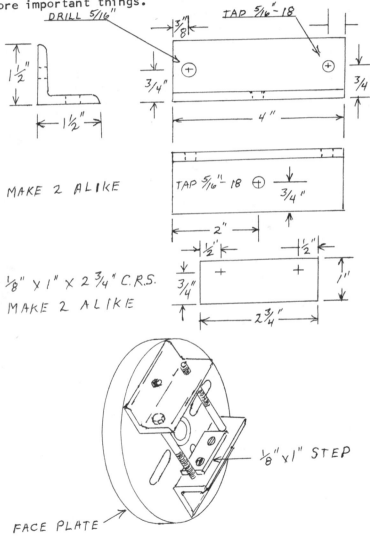

DRILL 5/16"

TAP 5/16"-18

3/8"

1 1/2"

3/4"

3/4"

1 1/2"

4"

MAKE 2 ALIKE

TAP 5/16"-18

3/4"

2"

1/2"

1/2"

1/8" x 1" x 2 3/4" C.R.S.
MAKE 2 ALIKE

3/4"

1"

2 3/4"

1/8" x 1" STEP

FACE PLATE

10

You can use standard structural angle iron, or you can make patterns and cast them in aluminum. A 1 1/2" angle, either 3/16" or 1/4" thick will work very well. If you do them in aluminum, make them 3/8" thick.

They won't be true square, so they need to be faced off to be accurate. Simply drill and tap them as shown on the drawings. Locate the holes carefully, so that both halves will be identical. Notice that the clamp bolts enter the clamp from opposite ends for balance. When the angles are brought face to face, one hole will be tapped, and the other is drilled.

To true them up, bolt one of the angles to the face plate with a 5/16" cap screw and flat washer. Test the angle between the leg and the face plate. Slip shims under either edge until the angle is exactly 90 degrees as indicated by an accurate try square.

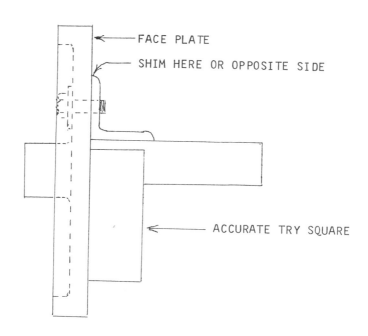

FACE PLATE

SHIM HERE OR OPPOSITE SIDE

ACCURATE TRY SQUARE

11

Now, bolt the mating angle to the leg of the mounted angle and face it off in the lathe.

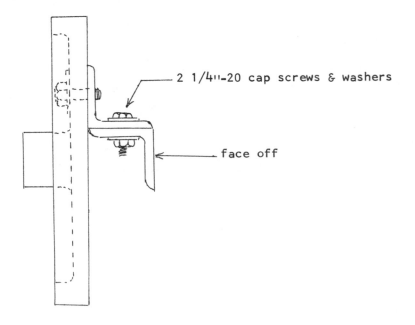

2 1/4"-20 cap screws & washers

face off

Finally, without separating the angles, invert the entire assembly. Mount the faced off leg on the face plate, without the shims, and face off the other leg.

If your face plate is true, and the original set-up was correct, both angles will be exact and the clamping surfaces will be parallel.

The 1/8" X 1" cold rolled flats are fastened to the inside of each angle to provide a step that is parallel to the surface of the face plate. This makes it possible to mount rectangular work so that you can face off two sides parallel. It also makes it possible to clamp rectangular work for accurate boring.

In use, the work is clamped loosely, the tail center is moved up to center it in the clamps, and the clamp bolts are tightened carefully to prevent shifting. Exactly as you do when you use a four jaw chuck. You will soon find dozens of uses for this simple fixture, and you will find ways to use the same principle on the work table of your shaper.

To improve the clamp further, carefully prepare a pair of vee blocks so that you can clamp short round work to do internal and external machining.

It can be tedious to form accurate vee blocks by hand. If you find it beyond your hand skill, simply mount cast aluminum blocks on the angles,and bore them to the size you need. You will soon acquire a set of useful sizes. When the shaper is complete, you can easily make accurate blocks.

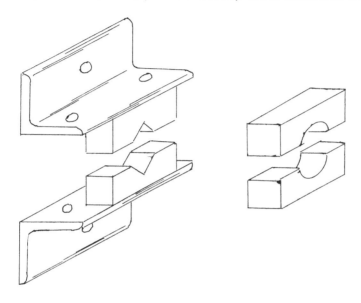

MAKE A SET OF ARBORS

The disc for the protractor, and the base for the head will be cast on a prepared arbor, so the means to mount it in the lathe is designed into the casting.

Other parts, such as the crank bearing support, will be mounted on arbors with a set screw to do diameter work and facing off.

It is impossible to drill a precise center in each end of a shaft, so you must begin with over-size stock to make your arbors. The diameter will then be concentric with the centers, and you can do accurate work on anything that you mount on the arbors.

Make your arbors as you need them, and save them for future jobs.

The sizes that are needed for this project are 3/8",
1/2", 5/8", and 3/4". I used a 3/4" X 8" arbor for machining the crank bearing support. For the rest of the jobs I
prepared a stepped arbor as illustrated.

Each step is 1 1/4" long, except the 3/4" step which is
2 1/2" long! File a flat on each step for the set screw to
seat on, so the work won't be damaged when you remove it.

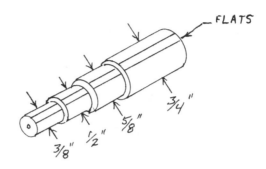

DRILLING AND TAPPING IN THE LATHE

You can order a 1/2" hand tighten chuck with a # 1 Morse
taper shank from Sears tool catalog at a very low cost. It
will fit the # 1 Morse taper socket in the head stock spindle, or the tail stock ram.

Its end is tapped for 1/4"-20 threads so that you can
use a draw bolt when it's mounted in the head stock. It is
not safe to use it in the head stock spindle without a draw
bolt, so don't take chances.

Its main use in the head stock spindle is as an arbor.
By removing the chuck and clamping work on the threaded end
with a nut and washer, you can turn disc shaped objects.

In this project, the main use for the chuck and arbor is
in the tail stock ram. You can do very accurate drilling
and tapping on work that is mounted on the face plate. You
can do the drilling under power, but the tapping is done by
rotating the work by hand as you feed the tap with the tail
stock.

There is no tang on the taper shank, so you must hold
the shank with a vise grip pliers so the taper shank and the
socket won't be damaged when drilling under power.

14

DRILLING A JOB WITH THE TAIL STOCK CHUCK

BORING A JOB WITH THE FACE PLATE CLAMPS

PATTERN MAKING ON THE LATHE

Some of the work can be done on the drill chuck arbor, and some can be done by fastening it to the face plate with screws. You can order a spur and cup center with # 1 Morse taper shanks from Sears tool catalog, which converts your metal lathe to a wood lathe. It's not safe to depend on a cone center to support wood, so be sure and get a cup center for wood turning.

You can machine the wood just as though it were metal. Very accurate patterns can be made using the carriage and automatic feed.

HAND WORK

A great portion of the work must still be done by hand, but when it is complete, the shaper will relieve you of much of this burden in the future.

Hacksawing, drilling, filing, and scraping are the main chores, but by this time you have built the lathe, and you will have a good foundation in these skills.

MAKE A SURFACE PLATE

The base of the ram and the vertical and horizontal slide supports are large surfaces that need to be scraped true.

A 24" aluminum level with milled surfaces makes a fair test standard, but a surface plate will be much better.

A commercially made surface plate is so expensive that its cost can hardly be justified for a hobby shop. You can make one out of plate glass that will be very accurate.

Plate glass is much better than sheet glass because it is very near true flat as manufactured. Sheet glass will require more work to true it up.

It requires a firm support over the entire area, because even 3/8" plate glass will bend or break when you apply any pressure.

An 8" X 16" plate will be large enough for this project, and most any job you are likely to tackle in the home shop.

If you have a table saw, or any other machine with a milled table, you can lay the plate glass on it to do your work.

To make a more accurate plate, you can lap the surface with grinding compound. For this you need two plates.

Mount the plates of 1/4" or 3/8" plate glass on bases made of plywood or particle board, so that they will be well supported.

Just make a level bed of plastic auto body putty on the board, and press the glass into the putty before it sets up. Bevel the edge of the putty so that it surrounds the glass.

Apply two or three coats of lacquer or varnish to the base so it won't take on moisture and warp.

When the putty has cured, smear valve grinding compound on the glass, and rub the two plates together in a back and forth circular motion. After about 200 strokes, you can separate the plates to see if they are uniformly frosted. Scrape off the compound and save it. Wash the plates clean, and dry them with a soft cloth. If there are any low spots, they will be unfrosted, and you will have to lap some more.

Accuracy depends upon firm support and even pressure. These plates will not be as perfect as their commercial counterpart, but they are adequate for the home shop.

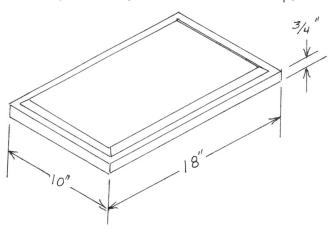

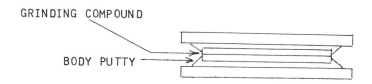

GRINDING COMPOUND

BODY PUTTY

You can lay a sheet of emery cloth on the surface plate to dress the surface of the work down in preparation for the scraping process. It will be faster and more accurate than filing.

The frosted surface of the plate is mildly abrasive, and it is very effective for testing the blued surface of the work.

The plate also provides an accurate base from which to do your vertical layout, such as locating the tail center for the ram.

DRILLING AND TAPPING

Step drilling is stressed throughout the project. It means to begin with a small pilot drill, and enlarge the hole with progressively larger bits. It is the only way you can be assured of accurate hole location and size.

On those members that require a row of screws, like the slide clamps or the column sides, the parts should be clamped together and drilled through. The outer member is enlarged to bolt size, and the inner member is tapped through the bolt hole. It's a bit tedious, but your bolt and tap holes won't line up unless you take this trouble.

FOUNDRY WORK

There are no really difficult castings in the project, but some of them are quite large compared to the lathe.

The first problem you will encounter, is that the sand will fall out of the cope when you attempt to close up the mold. This is called a " Drop Out ". Any other expression you might hear a molder make when this happens is likely to be unprintable, even though it expresses his true feelings at the moment. You can avoid the problem by adding ribs to the cope.

I use a 3/8" groove on the inside of my flasks, to grip the sand around the peremiter of the flask. I've made up several ribs from 3/4" stock, that fit the grooves to give extra support to the cope sand.

For really tricky molds, you can make up some "Gaggers" out of sheet metal. They fit over the ribs, like a saddle, to support the broad expanse of sand.

You will certainly need the ribs, and possibly one or more gaggers when you mold the column sides.

The dimensions will depend on the size of your flasks, and the problem at hand.

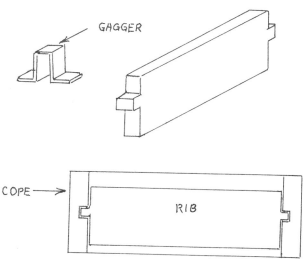

GAGGER

COPE→

RIB

SPLIT PATTERNS

It's not likely that you will use most of your patterns more than once or twice. For that reason, I strive to present them in the simplest form.

Flat sided patterns are no problem, because they are a simple matter to mold. Some shapes, though, have need for draft both above and below the parting line. While many of them can be molded by bedding, others need an odd side or a false cope or drag. Some shapes are so difficult to mold as a one piece loose pattern, that it makes sense to split the pattern at the parting plane.

By dividing the figure at the parting line, half will remain in the cope, and the other half in the drag. The two halves are aligned by free fitting pins so they will be easily separated when the mold is opened.

The pins are made of nails with the heads cut off, and they are installed in the cope half of the pattern. Holes are made in the drag half to register with the pins.

The drag half is layed on the molding board and rammed up. The drag is rolled over, and the cope half of the pattern is set in place. The cope is rammed up, a bottom board is rubbed in, and the cope is layed down to rap and draw the pattern.

The pins are easily installed if you use a nail of the same size to drill right through the cope half and into the drag half of the pattern. The holes are reamed slightly in the drag half, and the pins are cemented in the cope half.

They must fit closely enough to align the halves well, but they must fit freely enough so the halves won't bind up when you open the mold.

It's easy, and it's lots of fun. Some of the patterns are actually easier to make as split patterns, and they are certainly easier to mold.

BASIC STUFF

All of the castings for the shaper are aluminum. They are made in two part, closed, green sand molds. The one quart pot will hold enough metal to fill any of these molds.

I've had a lot of mail on the charcoal foundry, and some phone calls, all of which has been a great pleasure. There is a great willingness among shop enthusiasts to share the experience, and that adds much to the pleasure of the hobby.

Many have indicated an interest in a larger foundry, and some would like to try melting iron and brass. There is no reason why you can't do it, but I strongly advise gaining a good foundation of knowledge and experience first.

Remember that the total heat content of a charge will increase as well as the weight. Consider the problem of handling the molten charge before you build, not as an after thought. The radiant heat from a large pot can ignite your clothing if you are not well protected. There is not any time to think it over at the moment of pouring--you are likely to be forced to drop the pot and try to put out the fire. The consequence can be grim to say the least.

The danger is much greater with iron, because it pours at more than 2800 degrees. The pot is white hot, and can not easily be handled with ordinary tongs. I think it best to melt iron in a small batch type cupola, and pour with a shanked ladle rather than a crucible and tongs. Of course you must use a ceramic crucible rather than an iron or steel pot to melt iron.

Brass and bronze also pour at much greater temperatures than aluminum, and there are additional problems with flux and other technicalities. Most uninformed people seem to

20

believe that brass and bronze are easy to handle. The exact opposite is true; they are about the most difficult to master. Along with the problems of fluxing and temperature control, gates and risers are more technical. and patterns need much more study to bring the cavity in the right position for pouring.

My advice is: become a master at casting aluminum and pot metal before you tackle the " Hot Stuff ".

Metal casting is self instructing, because you simply can't make a casting without learning a great deal about patterns, molds, melting, and pouring. Your failures will be your best teacher, and you will have most of the material left to do the job over again the right way. The skill you gain will serve you in countless future jobs.

COMMON PROBLEMS

No one seems to be having any trouble with the basic principles of casting. Ramming too hard, or not hard enough is probably at the top of the list.

This is difficult to express without a demonstration, but you will quickly see your error in the casting. You will have gas bubbles and cavities on the surface of the casting if you ram too hard. The mold must be pourous so that gasses can escape through the sand. If you do not ram hard enough, the sand will yield at soft places, and there will be deformed places on the surface of the casting.

If you forget to vent with the wire, the result is the same as ramming too hard.

Gates and risers must be of adequate size to feed enough metal as the casting solidifies. Otherwise they will rob metal from the cavity and cause a srink cavity.

Supplies are the greatest problem, just as expected. It is very difficult to buy what you need in many areas. I was surprised to learn that fire clay could not be found in any city in the U. S..

Any of the pre-mixed castable refractories will serve to make a furnace lining, though they are quite expensive.

If you must blend your own molding sand, and you can't get fire clay, you can use Bentonite clay which is sold by farm supply dealers. It's used for lining the bottom of farm ponds to prevent seepage. Bentonite is very sticky; not much more than 5 % by weight will make good sand. It must be very thoroughly blended while the sand is dry.

Diatomaceous Earth is amorphous silica flour. It makes excellent parting dust, and was used for that purpose for a great many years. It has been replaced by high temperature plastic flour because it causes silicosis if you breath it. If you can't find parting dust, or a workable substitute, a bag of diatomaceous earth, as sold for use in swimming pool filters, will be enough for you and all of your friends. Be very sure to use protection to avoid breathing the stuff.

FUNDAMENTAL RULES

The essential elements of a pattern are draft and parting. Keep in mind that the pattern must be easy to draw as you make them up.

Ramming up a drag includes dusting with parting, ramming the sand, striking off, and venting with the wire.

A blank drag is rammed full and struck off to provide a smooth parting line for bedding a pattern.

Rolling over includes rubbing in the bottom board so it will give support to the sand when the cope is rammed up.

Ramming up a cope includes dusting with parting, setting sprues and risers, ramming the sand, striking off, and venting with the wire. The parting surface of the drag must be clean and smooth before the cope is rammed.

Sprues and risers are placed to feed the heaviest sections of the casting, and they must be of adequate size.

Swab the sand around the pattern with a soft wet brush, so the edge of the cavity won't break when you rap and draw the pattern.

Clean up the cavity and parting line carefully, and blow out the sprues and risers with the bellows. Turn the cope horizontal before you move it over the drag, so that loose particles will not fall into the cavity.

Aluminum pours best at about 1400 degrees fahrenheit. I stir the pot gently with an iron rod, to feel for lumps of unmelted stock or foreign material. If the red hot rod is clean, I know I am hot enough to pour. If there is a blob of aluminum on the end, it's too cool to pour.

Shrinkage and machining allowance has been considered, so don't make additional allowances on the patterns.

BOLTS AND NUTS

Cap screws are graded according to their hardness and

strength, and the grade is stamped on the head. The number of marks, plus 2, is the grade of the bolt. The largest grade number is 8, and that is the strongest bolt. Bolts that are not marked are grade 3, or less, and some of them are pretty poor stuff.

I used grade 5 cap screws on all of the slide clamps, because I didn't want to risk failure at these points. I used un-graded hardware on the remainder of the machine, because the graded stuff is much more costly.

Common flat head machine screws were used to mount all of the steel slides, and I see no signs of failure after taking heavy cuts on the table. Flat head cap screws are a better product, and those with the hex socket heads are even stronger.

Shoulder bolts, as used for the crank pin, are also called " Stripper Bolts ". They are made of alloy steel, and they are extremely tough and durable.

S.A.E. washers are smaller than U.S.S., and they fit the diameter of the bolt more closely. There are several spots where their use is indicated.

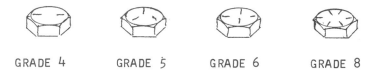

GRADE 4   GRADE 5   GRADE 6   GRADE 8

WE'RE READY TO BUILD

Assuming that you have established your foundry, and you have built the lathe, you will have no difficulty in building the shaper.

It's unlikely that you can complete any of these projects without losing some skin from your knuckles, or at least a bruise or two, but be extremely careful to avoid serious injury.

Follow each phase of the project at a liesurely pace, and enjoy yourself as you build your own metal working shop from scrap.

CHAPTER II

BUILDING THE COLUMN AND RAM

There are only three castings in the column. They are simple shapes, and the only problem is with the relatively broad surface of the cavity for the column sides.

THE COLUMN SIDE PATTERN

Both sides are identical, so only one pattern is needed. Remember that shrinkage and machining allowances have been considered in all of the drawings, so don't make additional allowances.

Take special care to make the pattern true square, so that the castings will mate well as opposing members. The rail that forms the ram slide ways is especially important.

This one can be made as a one piece pattern, but it is much easier to mold as a split pattern.

The main body of the pattern is rammed in the drag, and the small rail is rammed in the cope. Otherwise, you have to prepare a blank cope to bed the rail, and roll the mold twice if you use a one piece pattern.

Make the main body from a piece of 1/2" exterior glued plywood, add a 1/8" X 3/4" strip of pine at the top, and a 1/2" X 1" strip at the bottom. Fasten with brads and glue.

Provide generous draft on the inside of the 5 1/8" diameter hole, so the green sand core won't be damaged when you draw the pattern. Only minimum draft is needed on the rest of the pattern.

The rail that forms the ledge for the ram slide is a length of 1/2" quarter round, or base shoe molding. Locate the top edge carefully, so that it is 1/4" below the top of the pattern, and parallel to the bottom. If it is not parallel to the bottom, you will have extra work in fitting the ram slide.

Clamp the rail to the pattern, and drill four holes for the alignment pins, through the rail and into the pattern. Use a cut off nail as the drill, and make four short pins of the same size nail. The pins need only enter the drag half by about 1/4", and the ends should be smoothly rounded for easy entry. Cement them in the rail, (the cope half), and ream the holes in the drag half for a free fit with little side play.

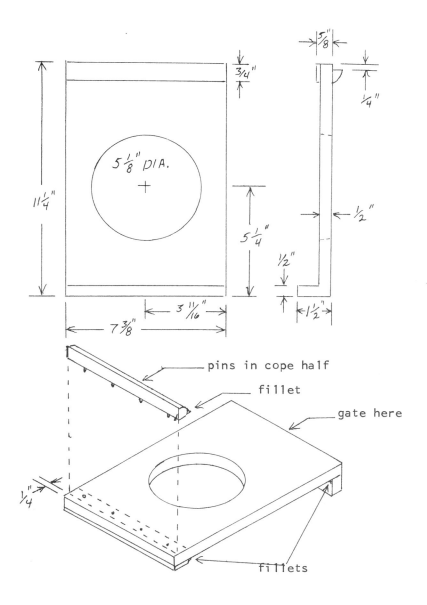

$5\frac{1}{8}''$ DIA.

$11\frac{1}{4}''$

$\frac{3}{4}''$

$5\frac{1}{4}''$

$3\frac{11}{16}''$

$7\frac{3}{8}''$

$\frac{5}{8}''$

$\frac{1}{4}''$

$\frac{1}{2}''$

$\frac{1}{2}''$

$1\frac{1}{2}''$

pins in cope half

fillet

gate here

$\frac{1}{4}''$

fillets

Wipe a small fillet of body putty at the inside corners on the drag half of the pattern.

Lay a sheet of waxed paper on the drag half, and set the rail in place with the pins registered in the holes. Wipe a small fillet at each inside junction, top and bottom, and allow the putty to set up. When the putty has set, remove the wax paper, re-assemble the parts and sand the fillet to mate well with the drag half of the pattern.

The fillet on the rail is made a part of the cope half so that the parting face of the drag is flat for easy molding.

Sand a minimum of draft to all outside surfaces, and slightly round all outside corners, except at the parting line.

Seal the pattern with two coats of lacquer or varnish.

MOLDING THE COLUMN SIDES

It will require a flask of about 12" X 16", and you will need about two ribs in the cope. If your sand is of good bond, you probably won't need any gaggers to help support the cope sand.

Lay the flat side of the main body of the pattern on a molding board, and ram up the drag. Give special attention to filling and ramming the sand in the hole. This leaves a green sand core that forms the hole in the casting, and it supports the cope sand when the mold is closed. You don't want any soft spots or voids in it.

Strike the drag off level, vent it with the wire, rub in a bottom board, and roll over.

Remove the molding board and blow off the parting face. Set the rail in place with its pins in the holes in the drag half of the pattern. Make sure that the holes are clean, and that the cope half of the pattern will separate easily from the drag half when you open the mold.

Dust the parting face with parting, set a 1" sprue pin about 1 1/2" away from the bottom edge of the pattern, and ram up the cope.

The ribs in the cope should divide the cope area into three sections, their bottom edge should clear the parting line by about 1/2". Peen the sand at an angle so that it will be forced under the rails, and be careful not to strike the rails as you fill and ram the cope in successive layers. If the rails are jarred it defeats their purpose, the sand body will crack, and you will have a drop out when you close up the mold.

Roofing nails can be driven into the sides of the rails, to give extra grip to the sand, or you can use sheet metal gaggers. A wet clay wash can be brushed on the ribs and gaggers if you have a serious bonding problem with your sand. Clay wash is made of fire clay and water, mixed to a soupy consistency.

Ram thoroughly and uniformly, but not so hard as to deform the drag sand. Vent generously with the wire.

Cut the top of the sprue to a funnel shape, and remove the sprue pin. Rub in a bottom board, and lift the cope straight up. Lay it on its back to swab, rap, and remove the pattern. Push a perfectly straight vent wire through the sand in the bottom of the cope cavity, so you will be sure of venting in that area.

Swab, rap, and draw the drag half of the pattern, and cut the gate from the sprue print to the cavity.

Clean up the cavities, blow out the sprue, and close up the mold. It requires a full pot of aluminum, and pour as fast as the sprue will accept the metal.

Let it cool for at least a half hour before you shake it out of the sand. An hour is even better if you can bear to wait.

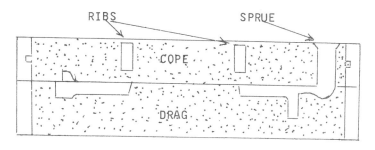

This is the routine sequence for all of the castings, so we won't have to discuss it in such detail unless there is a variation.

If you have difficulty with the cavity not filling, you may be pouring too cool, or you have not vented thoroughly. You could add risers at each end of the rail, so you can see the metal come up to indicate a full mold.

THE COLUMN FRONT PATTERN

This casting forms the front of the column body, as well as providing support for the vertical slide ways.

Like all of the slide way supports, the front portion is recessed to form raised pads so that there is less work to bring them true flat.

The back portion is recessed to provide clearance for the ram crank yoke.

The pattern is easiest to mold if it is split, though it can be molded as a one piece pattern.

It is actually made of three separate patterns, and it is aligned with two rows of pins.

The portion that forms the slide support is molded in the drag. It is an 8" length of 3/4" X 2 1/16" pine. There is a 1/8" X 1" recess between the 1/2" pads.

The cope half of the pattern is two 10 1/4" lengths of 3/4" X 1 1/8" pine. The " U " shaped opening at the top is 1 3/4" wide, and 2 1/4" deep. You can place a length of 1" wide stock between them to lay out the shape, and the same spacer will serve to hold the correct spacing when you drill through to install the alignment pins.

The critical dimension is the 3 1/4" finished width of the assembled pattern.

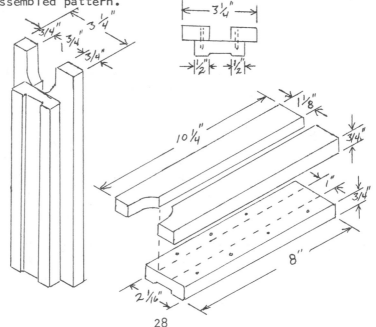

When the patterns are finished to dimension, assemble them with a sheet of waxed paper between them, and wipe a small fillet on each inside corner. When the body putty has set, remove the waxed paper and sand the fillets smooth.

Remember that the pins go in the cope half, and the drag half is smooth to lay on the molding board. The pins need enter the drag half just enough to provide good alignment, and the halves must separate freely when the mold is opened.

Provide minimum draft, and slightly round the outside corners except at the parting line.

Sand smooth all over, and seal with two coats of lacquer or varnish.

Molding is the same as for the column sides. Gate at the heavy end, and connect a riser to the end with a " U " shaped runner.

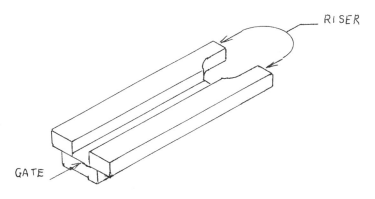

FINISHING THE COLUMN FRONT

The casting will shrink, and an amount of material will be removed as you file the sides parallel. Its finished width will be near 3 1/8". As you file the sides, bring them to right angles with the front surface. Perfection is not required here, just a reasonably straight surface that will mate with the inside of the column sides.

The most important job is to prepare the pads to accept the vertical slide ways. When the ways are installed, the column front becomes ˙a reference point for assembling the column sides.

INSTALL THE VERTICAL SLIDE WAYS

A 1/4" X 3" X 8" slab of cold rolled steel is fastened to the column front with 8 1/4"-20 X 3/4" flat head screws.

The surface of the pads must be worked true flat before the ways can be installed. This is a small job compared to the lathe bed, and it's done in exactly the same way.

You can do the job a lot faster if you lay a sheet of emery cloth on the surface plate, and work the pads to near flat by rubbing over the abrasive cloth.

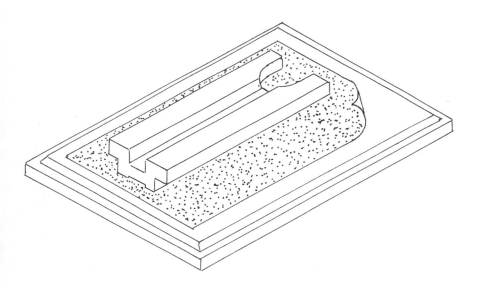

The surface plate backs up the emery cloth to present a level abrasive surface to plane off the drastic high spots.

When the pads are near flat, stain them with the prussian blue, test the surface on the cleaned surface plate, and proceed to hand scrape the high spots until they are true.

When the pads are true flat, center the steel ways so it overhangs equally on both sides of the casting, clamp it in position, and proceed to install the screws.

Step drill to tap size, right through the ways and the casting, enlarge the hole in the ways to bolt size, tap the casting through the bolt hole, countersink, and install the screw.

Install two screws, diagonally opposite, before you begin to drill the remaining holes. In this way, you can be sure that your work won't slip out of alignment.

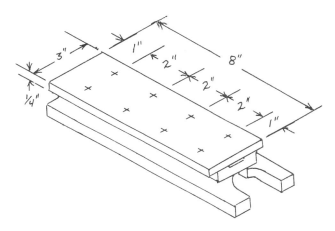

Center the rows of screws so that they fall 1/4" inside the edge of the pads. Countersink deeply enough so that no part of the screw head is above the surface of the ways, but not so deep as to weaken the assembly.

ASSEMBLE THE COLUMN

The ram slide channel must be finished before you can align the parts fo the column, but you can begin assembly now.

The vertical slide ways must be aligned at exact right angles to the ram slide channel, so only the two top bolts are installed at this time.

Each column side is prepared by drilling two 5/16" holes in the mounting feet, 1 3/4" from each end.

A 3/8" hole is drilled 1" above the mounting foot, and 1/2" from the back edge of the side. A length of 3/8"-16 threaded rod is installed as a spreader, with nuts, washers, and lock washers, both inside and outside.

A row of 1/8" pilot holes is drilled 3/8" from the front edge, and on 2 5/8" centers, beginning with 1" above the mounting foot.

Remember that there is a right and left hand member, so lay them out as opposing sides.

Clamp the column front casting between the sides, and align it as near true square to the sides and the ram channel as you possibly can.

Install the spreader bolt at the rear bottom, and adjust it to bring the sides parallel.

Bolt the mounting feet to a piece of plywood, and check the alignment again.

Extend one top pilot hole through the side, and through the column front casting.

Enlarge the hole by steps to 1/4", then enlarge the hole in the column side to 5/16". Be careful not to run the 5/16" drill into the column front casting.

Tap the hole in the column front for 5/16"-18 threads, and use the bolt hole in the column side to guide the tap.

Install a 5/16"-18 X 1" cap screw with a lock washer, and re-check the alignment of the parts before you begin to install the other top bolt.

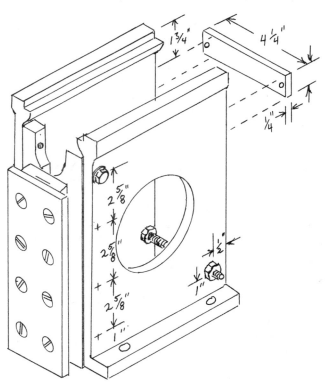

When the second top bolt is installed, measure the width across the front of the column to determine the length of the top rear spreader. It should be 4 1/4" or less, and the ram slide channel should measure not more than 3 1/4".

The ram slide is 3" wide, and the gib is 1/8" wide, so the channel needs to be at least 3 1/8" wide when it has been scraped to fit the ram. If it is any less than 3 1/8" at this stage, add shims between the column front and the sides to widen it.

When the width of the channel is established, make the top rear spreader to fit, and install it 1 3/4" from the top edge of the column side with two 1/4"-20 X 1" cap screws and lock washers. The spreader is a length of 1/4" X 1" cold rolled steel. Clamp the spreader in place, step drill 3/16" holes through the spreader and into the castings. Enlarge the holes in the spreader to 1/4", and tap the holes in the casting, using the spreader hole to guide the tap.

You will be tapping blind holes, so drill them about 1" deep, and tap about 3/4" deep. It is very easy to break a tap when tapping blind holes.

The finished ram is needed to finish the slide channel so that the vertical slide ways can be adjusted to exact right angles before the other six side bolts are installed.

Set the column aside for now, and begin on the ram.

THE RAM PATTERN

Since you have no way to bore the end of the ram for the tool head pivot, it is cast with a steel core to provide a bore.

It would be impossible to remove such a steel core, so the end of the ram is made with a half bore, and a cap is bolted on to make the full bore.

The steel core is used in both the ram and the cap, and it is also used to provide a center for mounting the ram in the lathe for facing off. You need to make the core from a length of oversize stock, so the center will be concentric.

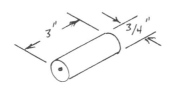

The base of the pattern is a 12 1/4" length of pine, 2 1/8" wide, and 3/4" thick. A 2 1/8" X 3/4" X 3" block is glued to the top at one end, and the 1 1/2" X 3/4" X 8 1/4" rib is glued in place. Small nails will hold the parts as the glue dries, and they can be left in permanently.

The print for the steel core is a 3" length of 3/4" dowel rod that is split exactly in half. Half of the piece will be waste, because of the width of the saw cut. The half you use should be as nearly exact as you can make it, so that the core print will not be oval shaped.

Position the core print in the center of the pattern, and as near to parallel to the ram sides as possible. Glue and nail it in position.

The bottom surface of the pattern is recessed 1/8" deep, to leave 1/2" wide pads around the perimeter of the base.

The area indicated by the dotted lines is filled with plastic auto body putty, and faired out smoothly. It takes several applications of the putty in order to end up with a smooth surface. You simply can't sculpture so large a mass of putty in one application.

Provide minimum draft on all vertical surfaces, and sand all outside corners round, except at the parting plane.

MOLDING THE RAM

Simply lay the base of the pattern on the molding board and ram up the drag.

Vent very generously with the wire, rub in the bottom board, and roll over the drag.

Clean up the face of the drag, especially the recess in the base of the pattern, and set a 1" sprue pin on each side of the pattern, about 1" away from the heaviest part.

Dust with parting, and ram up and vent the cope.

Swab all around the pattern, rap and remove it, and cut full size gates from the sprue prints to the cavity.

If your sprue is undersize, or the runner is too small, you will have a serious shrink cavity.

Smoke the steel core over a candle flame, and fill the centers with graphite so they won't fill with aluminum.

Set the core in its print, close up the mold, and pour as fast as it will accept the metal.

Give this one at least an hour to cool before you shake it out.

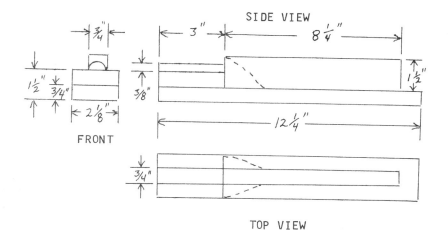

SIDE VIEW

FRONT

TOP VIEW

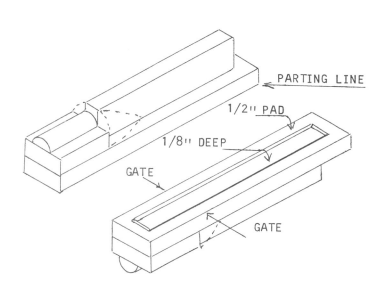

PARTING LINE

1/2" PAD

1/8" DEEP

GATE

GATE

35

THE RAM CAP

A very simple casting which can be poured with a single pop gate at the center.

The raised pad in the center locates the sprue pin, and provides a fillet at the junction of the pin and cap.

The core print is another split 3/4" dowel, and the same steel core is used in the mold.

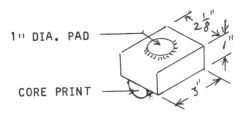

1" DIA. PAD

CORE PRINT

MOLDING THE CAP

Prepare a blank drag, and bed the core print in it.   The remainder of the pattern will be in the cope.

Set a 1" sprue pin on the raised pad, and ram up and vent the cope.

Pull out the sprue pin and rub in a bottom board.

Open the mold and lay the cope down to swab, rap, and remove the pattern.

The same steel core that was used for the ram is prepared in the same manner and set in the print.

Close up and pour rapidly.

The sprue is cut off about 1/4" above the casting, to leave a raised boss for the set screw.

FINISHING THE RAM

Because of shrinkage the core fits very tightly in the half bores.   You will have to drive it out with a hammer and punch, but be careful not to damage the lathe center.

An amount of filing and scraping is needed before the cap and ram fit properly, and the core fits the bore without play.

It will be necessary to file the joint, and possibly to install shims in the joint to get a proper fit on the core. Ultimately the 3/4" arbor of the tool head will fit in the bore, and it must be firmly supported.

Use a clean 3/4" arbor to fit the cap to the ram. Clamp the cap in place, with the arbor in the bore, and add any necessary shims to bring it to a snug even fit.

Step drill four 1/4" holes through the cap and through the ram. Enlarge the holes in the cap to 5/16", and tap the holes in the ram for 5/16"-18 threads. Install four 5/16" X 1 1/2" cap screws with lock washers.

Remove the arbor, and step drill a 5/16" hole in the top of the cap, centered in the raised boss, and as near to the center of the bore as you can manage. Tap the hole for 3/8"-16 threads. This will be the set screw hole for the tool head arbor.

FACE OFF THE RAM

The ram is mounted between centers to face off the front end, and for this you need to drill a 60 degree center hole in the tail end of the ram and install the centered core in the front end.

The base of the ram must be brought to true flat before you can accurately locate the tail center height. Use the same method as for the vertical slide support, and take extra pains to be sure the base is true flat.

Use the surface plate or other true flat surface to lay out the tail center.

Measure the exact height of the front center, and transfer it to the tail end of the ram.

Drill the 60 degree center hole in the tail end of the ram, and mount the ram between centers in the lathe. The front end of the ram will be supported by the lathe tail stock, and you can drive the work with the face plate angle clamps.

The rib on the ram has been shortened by 1" in order to bring its length within the capacity of the lathe. If you have a clearance problem with the tail stock ram extended about 1 1/4", you can extend the arbor in its bore, or you can use a longer arbor in order to have clearance to operate the cross slide for facing off.

On my lathe, I had to remove the ball crank from the lead screw and clamp the tail stock about 1" beyond the end of the bed in order to gain the clearance.

Make enough passes with the cross slide to clean up the front surface of the ram so that it will be at right angles to the center line.

This operation provides a firm foundation for mounting the base disc for the rotating tool head.

You can't see the head stock center in the photograph, but it's there. The face plate angle clamps are off center, and they're tightened just enough to hold the ram. Be sure you don't force the ram off center when you tighten the clamp.

INSTALL THE RAM SLIDE

This is a 12" length of 1/4" X 3" cold rolled steel. It is fastened to the ram with twelve 1/4"-20 X 3/4" flat head screws at 2" intervals.

The ram will be slightly off center on the slide because the left hand clamp covers more of the slide than the right hand clamp.

Position the ram on the slide 1/2" from the left hand edge of the slide, and clamp it securely as you step drill the holes and tap them. Like the vertical slide ways, put two screws diagonally opposite, to prevent slipping, and follow the same procedure for drilling, tapping, and countersinking the holes.

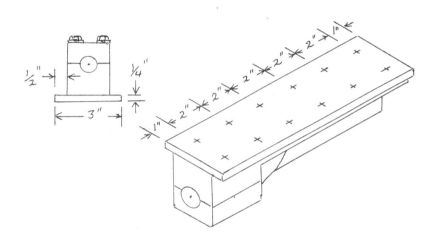

The steel slide should be a tiny bit behind the front of the ram so it doesn't interfere with the tool head base.

The finished ram will provide the test standard for fitting the channel, which is one of the more demanding chores in the project.

When the ram channel is finished, it will be the reference point from which the column is finished.

FITTING THE RAM CHANNEL

You can wrap a sheet of emery cloth over the ram slide, and use it as a tool for rapidly removing drastic high spots from the channel. Simply draw it back and forth in the channel, and clean it up for scraping.

A diagonal groove should be cut in each corner of the channel, to separate the horizontal and vertical pads.

If you failed to position the rails properly on the pattern, you will have high corners in the channel, and these must be brought even before you can do any accurate finishing. The ram must rest firmly in the channel without any rocking, or it won't be effective as a test standard. Just file and scrape the high corners down until the ram rests in the channel without rocking.

From this point, it's a simple matter to blue the channel, test with the ram, and scrape down the high spots.

The left vertical pad must be scraped true, but the right side need not be done so carefully because the ram will bear on the gib.

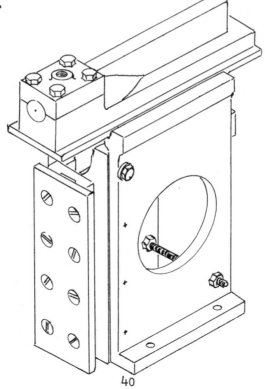

When the ram channel has been scraped to at least 75% contact with the ram, you can file cown the clamp pads to bring them parallel to the bottom wear surface of the channel. A short length of 1/4" X 3" cold rolled steel will serve as a gauge to prevent filing too much. When the file touches the gauge, you have a 1/4" depth. Once the clamp pads are brought parallel to the wear surface of the channel, and you have an even channel depth of 1/4", you can carefully file off an additional several thousandths of an inch so that shims can be installed with the clamps.

COMPLETE THE COLUMN BODY

Now that you have established the ram slide channel, it will be easy to adjust the vertical slide to exactly 90 degrees to it.

Use an accurate try square to adjust the angle, clamp the column front securely, and finish installing the side bolts. Drill through the 1/8" pilot holes, and all the way through the column front casting. Enlarge the holes to 1/4" all the way through, and enlarge the holes in the column sides to 5/16". Again, be careful not to enter the column front casting with the 5/16" drill. Tap the holes in the column front for 5/16"-18 threads, and install the bolts.

INSTALL THE RAM SLIDE CLAMPS

Two 7 1/4" lengths of 1/4" X 1" cold rolled steel, and a 7 1/4" length of 1/8" X 1/4" key stock will complete the ram installation.

The clamps are each fastened with four 1/4"-20 X 1" cap screws.

There are five # 10-24 gib screws to adjust the fit of the ram in the channel.

The clamps are installed with several shims in each side, to provide for take up after wear.

Notice that the right hand clamp is set in from the outside edge in order to have as much clamp as possible on the gib side. The center of the row of screws will be slightly different than the left side.

Install both clamps with the ram in place, so you will be certain of clearance between the clamp and the ram casting. Be sure that the ram is bearing on the left side before you drill any holes.

41

The gib screw holes must be drilled and tapped carefully. The end of the screw is ground to a point which seats
in a dimple in the gib that is made with the tap size drill
as you drill through the column side. Install one screw
completely before you drill the rest of the tap holes so it
doesn't slip out of alignment. Install a jamb nut on each
gib screw.

Heavy duty aluminum foil will make good shim stock for
the clamps. The shims should be from .001" to .002" thick,
and there should be several in each side so you can adjust
the clamps after wear in use.

The drawings will show you the spacing of the bolts and
gib screws. Step drill the clamp bolt holes through the
clamps and about 1" into the side casting. Tap them about
3/4" deep, and be extra careful tapping these blind holes.

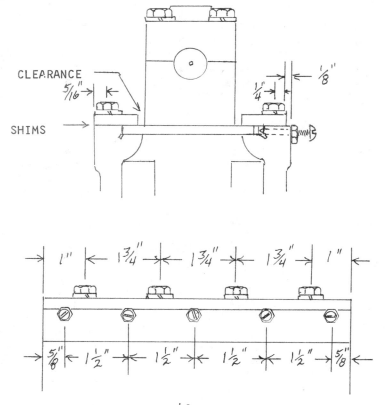

CHAPTER III

THE RAM CRANK MECHANISM

A simple scotch yoke mechanism, using the sliding block in a yoke to convert rotary motion to reciprocating motion.

Since the lower pivot is anchored, the upper end of the yoke is forced to move back and forth as the crank pin slides the block through the yoke.

The crank pin is adjustable in the slotted crank plate so that the length of the stroke can be varied.

The ram clamp is also adjustable so that you can set the beginning of the stroke properly.

On commercially built shapers, the " Bull Wheel " is a large gear driven by a pinion. I chose to use a standard chain sprocket because they are easy to find, and much less costly than gears.

Even at todays prices, I only had to lay out about $15.00 for both sprockets. Considering that the type of machine we're building would cost about $3000.00, if we could find one, I thought I had a bargain.

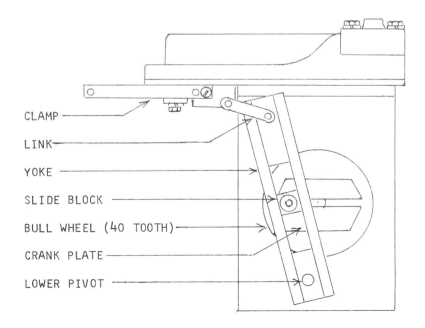

CLAMP

LINK

YOKE

SLIDE BLOCK

BULL WHEEL (40 TOOTH)

CRANK PLATE

LOWER PIVOT

THE BULL WHEEL AND PINION

Standard, finished-bore, sprockets in a number 35 size. The bull wheel is a 40 tooth with a 3/4" bore, and the pinion is a 10 tooth with a 5/8" bore.

The pinion is used as is, but the bull wheel must be mounted on an arbor and faced off to prepare it for mounting the crank plate. These sprockets are made with a heavy hub welded to the tooth plate, and the surface must be faced off on the lathe. Be careful not to damage the teeth when you face it off. Remove only enough material to clean up.

The crank plate is built up of two pairs of 1/4" X 1" cold rolled steel plates to form a 1/4" X 5/8" Tee slot on the face of the sprocket.

The assembly is carefully centered on the sprocket, and fastened with four 5/16"-18 X 3/4" grade 5 cap screws.

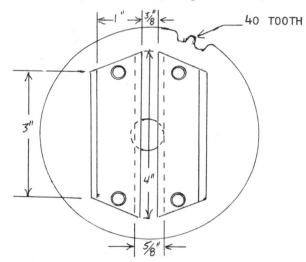

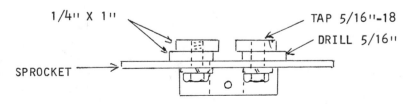

44

As in all stacked assemblies, clamp the members securely, and step drill to tap size through all members.   Only the outside plate is tapped.   The sprocket and the inside plate are drilled to 5/16" bolt size.

The standard hub on a 40 tooth sprocket should be 2 1/4", though you may get one with a 2 1/2" hub.   If necessary, mount it on an arbor and reduce the diameter of the hub to make room for the bolt heads.

The crank pin is a 1/2" X 3/4" shoulder bolt.   These are also called " stripper bolts ".   It has a 1/2" diameter on the shoulder, the threaded portion is 3/8"-16, and the head is round with a hex socket.

The sliding nut is made from a 5/8" length of 1/4" X 1" cold rolled steel.   It must be filed to a free sliding fit in the crank plate Tee slot, and it is tapped 3/8"-16 in the center.

The threaded portion of the bolt may have to be cut off at final assembly, but don't shorten it yet.

One or more 3/8" S.A.E.   washers are used to align the slide block with the yoke.   The spacing needs to be found at final assembly, so just set it aside for now.

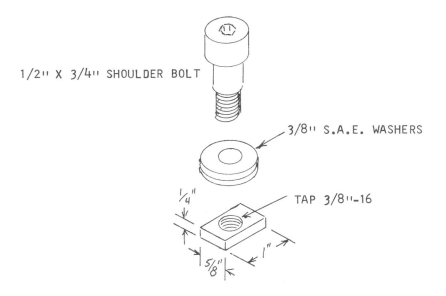

1/2" X 3/4" SHOULDER BOLT

3/8" S.A.E. WASHERS

TAP 3/8"-16

1/4"

5/8"

1"

THE CRANK BEARING SUPPORT

This is a simple figure that is a pure delight to cast and machine. Making it really proves the worth of the lathe and the charcoal foundry.

THE PATTERN

A 6 1/2" disc of 1/2" plywood, with a 2 1/4" diameter by 2" long hub is fastened together with glue and nails. Mount it in the lathe, either on the face plate or between centers, and reduce the large diameter to 6 1/4", and the hub to 2".

Sand a minimum of draft before you take it off the lathe, and seal with two coats of varnish or shellac.

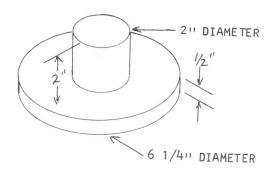

MOLDING THE BEARING SUPPORT

This would be an easy figure to ram up in the drag, but it requires a large riser to feed the hub, and it would be more difficult to cut off the riser.

Prepare a blank drag, and rest the base of the pattern on the parting face.

Set a 1" sprue pin to feed the edge of the disc, and a 1 1/2" riser in the center of the hub.

The 1 1/2" riser will serve as a sinking head for the hub, and it will be easy to cut off.

Ram up and vent the cope, and pull the sprue and riser before you rub in the bottom board.

Lay the cope down to swab, rap, and remove the pattern.

Cut the gate in the cope, and form a fillet at the base of the riser with a wet swab.

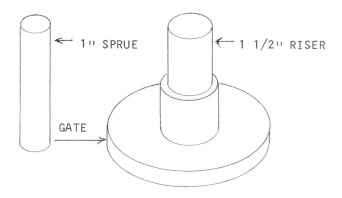

BORING THE BEARING SUPPORT

The center of the hub is bored to .999" to accept a pair of 1" X 3/4" flanged bronze bushings.

3/4" I. D. ——→

1" O. D.

If you have difficulty locating the bushings, you can get Dayton pillow blocks, stock # 2X530, and drive the bushings out. They retail at about $4.00, and they are available from W. W. Grainger Co., who has outlets all over the U.S..
I used Dayton pillow blocks for the countershaft and the pinion shaft in the 5/8" size, which is stock # 2X529.
You could use straight 1" X 3/4" bushings for the crank bearings, with standard machinery bushings for thrust.

MOUNT THE CASTING ON THE FACE PLATE

Drill two 1/4" holes in the flange of the casting, and tap them for 5/16"-18 threads.
Drill a 3/8" hole in the center of the hub, and mount the casting on the face plate with two 3/8" set screw collars for spacers. The 3/8" hole permits entry for the boring bar, and the spacers allow it to pass through without running into the face plate.

Of course the live center is driven out for this operation, and you should stuff a small bit of rag in the bore
to keep it clean.

It takes considerable practice to bore an accurate hole
with a small forged boring tool. You must take light cuts,
and measure frequently so that you can predict the finished
size on the last pass. The thin shank of the tool tends to
spring away from the work, even on very light cuts.

The bore wants to finish at .999", which is just .001"
less than 1", so that the bushings will be a forced fit. If
it is too small, the inside bore of the bushing will be dis-
torted, and you will have a reaming job on your hands.

Take your time, and be very careful, but if you happen
to bore oversize, you have enough hub stock to install a
sleeve and try again.

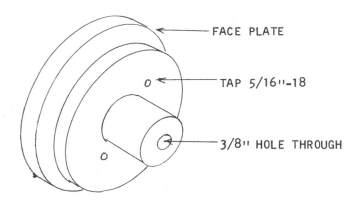

FACE PLATE

TAP 5/16"-18

3/8" HOLE THROUGH

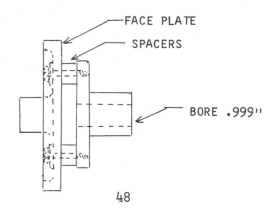

FACE PLATE

SPACERS

BORE .999"

Exotic measuring equipment is not required for this, or any other part of the project. I used a $13.00 vernier calipers to build my shaper, and found no need for anything better. It is a convenience to be able to read in divisions of .001", and you can get inexpensive vernier calipers that will do the job.

If you must limit the amount of money you spend on tools, I suggest a 6" vernier calipers, rather than a cheap dial calipers or micrometer. Particularly the imported plastic dial calipers, which offer much, but are a real disappointment.

A stainless steel vernier calipers will perform the job of an entire set of micrometers, for both inside and outside measuring, and it takes only a little practice to master the tool to produce very precise work.

To make inside measuring easier and more accurate, you can prepare a test gauge out of cold rolled steel. For the job at hand, just machine a cylinder with three steps so it can be used to test the bore when you approach the final cut.

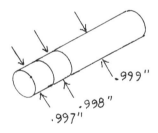

.999"

.998"

.997"

When you have finished boring the bearing support, face off the end of the hub to clean it up. The finished length of the hub is not critical, it can be from 2 1/4" to 2 1/2" through the entire length of the bore.

Install the bushings, and drill and tap a 1/4"-20 set screw hole in the hub.

The set screw will lock the casting on a 3/4" arbor for the outside machining, and it's later drilled to 1/4" for a spring cap oiler.

The arbor should be about 8" long, and grind a flat for the set screw to bear against so the bore won't be damaged when you remove it.

Mount the arbor between centers and face off both sides of the disc just enough to make them parallel.

Machine the hub to 1 3/4" diameter for about 3/4" of its length.

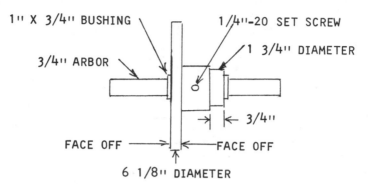

Machine the edge of the flange to make it smooth and concentric, but don't make it smaller than 6 1/8" in diameter.

There will be a space in the bore that is not covered by the bushings. This will be the oil reservoir for the crank bearings. When you have finished the lathe work, enlarge the set screw hole to 1/4", and install a spring cap oiler in the hole. A piece of felt, about 1/8" thick, is put in between the bearings to soak up oil and wipe the center of the drank shaft.

INSTALL THE CRANK BEARING SUPPORT

The flange is fastened to the right side of the column with four 5/16"-18 X 1" cap screws with lock washers.

Position the support casting so it is centered over the 5" hole in the right column side. Make sure that the oil cap is in the up position, and clamp it in place.

The bolts are centered on a square pattern, about 4" between centers. Make sure that the holes will enter the column side casting, and that they are centered about 5/16" in from the edge of the flange.

Step drill four 1/4" holes, enlarge the holes in the flange to 5/16", and tap the holes in the side casting for 5/16"-18 threads.

Install four 5/16"-18 X 1" cap screws with lock washers.

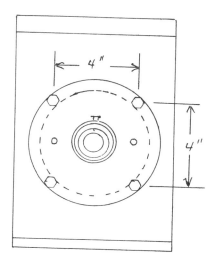

INSTALL THE CRANK

The length of the 3/4" crank shaft will be about 4", but you will need to determine the exact length when the feed crank is installed. For now, install the bull wheel on a short shaft, and use a set screw collar on the outboard end of the shaft.

INSTALL THE PINION BEARINGS

I used Dayton pillow blocks, stock # 2X529, though any similar pillow block will do fine. These are 5/8', self aligning, bronze bearings. I discarded the rubber pads that were furnished with the bearings.

Two angle iron adapters are bolted to the column side,

using slotted holes to provide for adjusting the chain tension.

The slotted holes are formed by step drilling 5/16" holes, and cutting from the edge with a hack saw.

The location for the pillow block mounting holes must be taken from the blocks you use. Mount them as near to the bottom of the adapter as possible.

Remember that there is a right and left hand member.

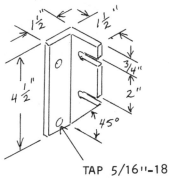

TAP 5/16"-18

The adapters are positioned to rest against the ledge at the top of the column side casting. Locate hole centers to match the slots, step drill them to 1/4", and tap for 5/16"-18 threads. Install the brackets with 5/16"-18 X 3/4" cap screws with flat washers and lock washers.

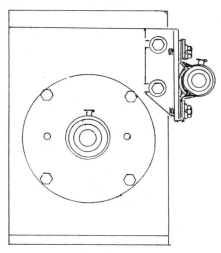

Install the pillow blocks with 5/16"-18 X 3/4" cap screws, and leave them loose for now.

Loosely assemble the 10 tooth pinion sprocket and two set screw collars on a 5/8" shaft, and install them in the pillow blocks.

The chain will be something less than 24". You can determine its actual length from a trial assembly, remove unneeded links, and join with a repair link.

You can buy a 24" length of # 35 chain from most farm supply dealers or industrial supply houses.

The slotted holes in the pillow blocks and the adapter brackets will permit aligning the shaft and adjusting the chain tension. The chain should fit with a slight sag.

Install a 4" - 3 3/8" - 2 5/8" - 2" step cone pulley on the left end of the shaft.

Now you can align all of the chain drive components, and lock the set screws lightly.

Disassemble and cut the shaft to its finished length, which is about 10 1/4". Grind flats for the set screws to seat against to protect the bores of the components.

PINION

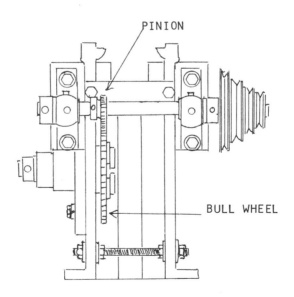

BULL WHEEL

THE CRANK YOKE

Two 10" lengths of 1/2" steel key stock, two small cast-ings, two bronze bushings, and four cap screws make up the crank yoke.

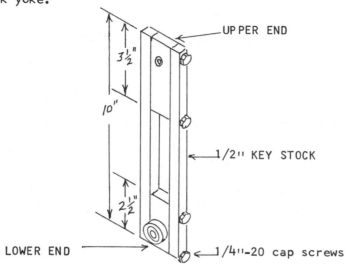

UPPER END

$3\frac{1}{2}$"

10"

1/2" KEY STOCK

$2\frac{1}{2}$"

LOWER END

1/4"-20 cap screws

THE END CASTINGS

Both castings are essentially rectangles, and though the lower end has a boss protruding on each side, it isn't worth the work to make a detailed pattern.

Just make two blocks with slight draft, and ram them in the same mold. Feed them both with the same 1" sprue.

These castings provide the rough stock for machining on the lathe, and it's an interesting job.

The larger casting also provides the stock for the slide block.

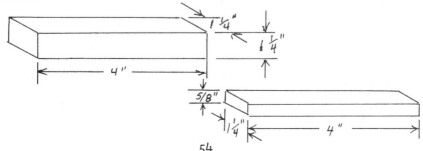

$1\frac{1}{4}$"

$1\frac{1}{4}$"

4"

5/8"

$1\frac{1}{4}$"

4"

MACHINE THE LOWER END

This is a simple series of operations, and when it is done you will have difficulty in convincing some people that it was done on a simple lathe.

The face plate angle clamps will hold the stock, and you have facing off, tailstock drilling, and boring to do.

Begin with a 2 5/8" length of the 1 1/4" square block, and face off both ends to bring it to 2 1/2" long.

Mount the stock in the clamps and face off the first side until it is clean.

Turn it over and face off the opposite side until it is exactly 1" thick. The 1/8" X 1" steps in the clamp will register the work so that the surfaces will be parallel if you have installed them carefully.

Face off the third side until it is clean, and turn it over to face off the fourth side, until you have a 1" true square block that is 2 1/2" long.

DRILLING THE THROUGH BOLT HOLES

If you have a drill press and a good vise for the table, this is a simple job, but don't try to drill these holes in a free hand manner.

A pair of 1/4"-20 X 2" cap screws must pass through each end to hold the yoke together. It is very near impossible to do the job free hand.

Remove the faced off stock from the lathe, and carefully mark and center punch the hole locations.

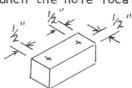

Clamp the block loosely in the face plate clamps, and bring the tail stock center up, to position either center on the turning center of the lathe. Tighten the face plate clamp carefully, as you hold the work with the tail stock center in the punched center mark of the hole.

Install the tail stock chuck, and drill a 1/8" hole in the block. Remember to hold the shank of the chuck adapter as you drill.

Enlarge the hole with a 3/16" bit, and do the other hole in the same manner.

These 3/16" holes first serve as a guide for drilling the holes in the steel key stock before they are enlarged to 1/4".

BORING THE LOWER END

The lower end is supported on a 1/2" shaft, and it has a 5/8" X 1/2" bronze bushing.

Carefully mark and punch the center of the bore on the block, and center it in the face plate clamps with the tail stock center.

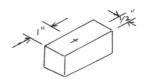

Step drill a 3/8" hole, and carefully enlarge it with a boring bar to .624", which is .001" smaller than 5/8".

Remove the block from the clamp and install a 5/8" X 1/2" bushing in the bore.

Install the head stock center, slip the bore onto a 1/2" mandrel, and mount it between centers. Use the clamps to hold the casting, as you face it off to leave a 1/4" high boss on each side of the lower end casting.

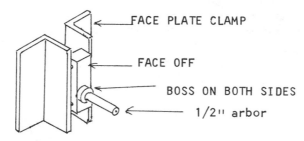

A 1/8" oil hole is drilled diagonally to enter the center of the bushing. Scrape the burr off the inside of the bushing after you drill the hole.

This all may seem a little tedious, but the work must be done accurately, and it will be worth the effort.

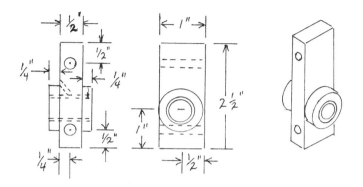

YOKE LOWER END WITH 5/8" X 1/2" BUSHING

THE YOKE UPPER END

Trim the 1 1/4" X 5/8" stock to 3 1/2" long, and face off all four sides to reduce it to exactly 1/2" X 1".

The 3/16" Through holes are drilled in the same manner as the upper end.

The link pivot hole is bored to .374", to provide a fit for a 3/8" X 1/4" bronze bushing.

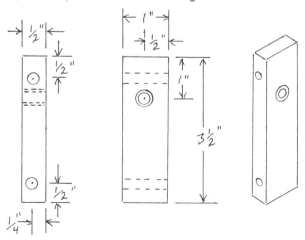

THE SLIDE BLOCK

The remaining stock from the lower end casting is faced off to 1" true square, and 3/4" thick.

A .624" hole is bored through its center to provide a press fit for a 5/8" X 1/2" bronze bushing.

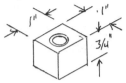

ASSEMBLE THE YOKE

Use the upper and lower ends as a guide for drilling the holes in the steel bars, and you will have no trouble with aligning the parts for assembly.

Clamp the ends to one of the legs, and drill four 3/16" holes. Be sure that the parts are well aligned and securely clamped.

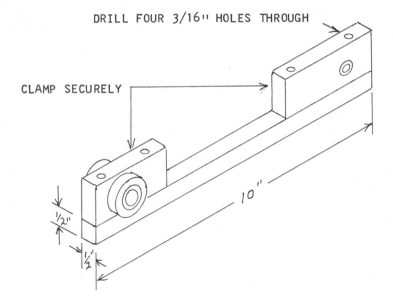

Slip four 3/16" bolts through the drilled leg, and into the castings, so that the members won't slip as you clamp the other leg in place.

Invert the assembly and drill the 3/16" holes through the other leg.

Enlarge all four holes to 13/64", and mark the parts so you can re-assemble them as they were drilled.

Enlarge four holes in one leg to 1/4".

Enlarge the through holes in both castings to 1/4".

Re-assemble the parts in the original drilling position, and tap 1/4"-20 threads, using the through holes to guide the tap. There will be considerable strain on the tap, so be very careful as you tap these holes.

Assemble the yoke with four 1/4"-20 X 2 1/4" cap screws with lock washers.

The bolt heads will face the front of the machine, and the oil hole in the lower end will be on the left side.

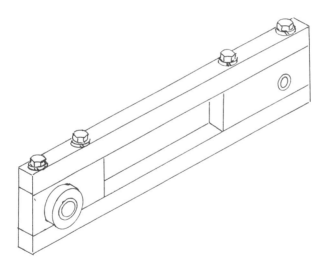

Cut off any portion of the through bolts that extend beyond the opposite leg.

THE YOKE SUPPORT BOSSES

Two are required. They are simple castings, and fun to machine on the lathe.

The pattern is a rectangle of wood, and the 1 1/4" sprue forms a boss on the casting.

1 1/4" SPRUE PIN ──────────→

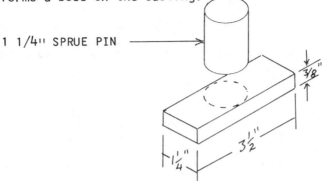

Ram the pattern into a drag. When you roll it over, set the 1 1/4" sprue pin in the center of the pattern and ram up the cope.

When you clean up the mold, form a fillet at the base of the sprue with a wet swab, so you won't have a shrink flaw at the junction.

MACHINE THE BOSSES

Cut the sprue off 7/8" above the base, and mark and punch the center of the round.

Mount it on the face plate clamps and align the center with the tail stock center.

Step drill a 3/8" hole with the tail stock chuck, and finish it to exactly 1/2" with a boring bar.

Drill and tap a 1/4"-20 set screw hole in the side of the casting, and lock it on a 1/2" arbor.

Mount the arbor between centers and face off both sides and reduce its length to 1".

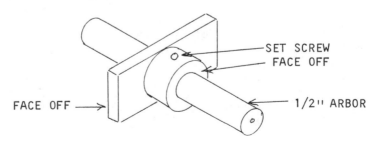

SET SCREW
FACE OFF
FACE OFF ──→
1/2" ARBOR

INSTALLING THE CRANK YOKE

The bosses are fastened to the inside of the column by four 5/16"-18 X 1" cap screws with lock washers.

The exact location will have to be determined by a trial assembly.

Slip a 3" length of 1/2" steel shaft through the lower bushing of the yoke, and slip a boss on each end. The shaft should have a flat ground on each end for the set screw to seat against.

The heads of the through bolts will face to the front, and the boss set screws face upwards. The oil hole in the yoke lower end is on the left.

Tighten the set screw in the right hand boss, but leave the left hand set screw loose.

Slip the yoke assembly through the rear opening in the column, and loosely clamp the bosses to position the lower pivot center at 1 3/4" above the lower edge of the side, and 3 1/2" forward of the rear edge of the side. This is just an approximate position, you will need to operate the crank by hand to find the exact position.

Install the slide block and crank pin with enough 3/8" S.A.E. washers to position the slide block about mid-way in the yoke. If the threaded portion of the crank pin is too long, cut off the un-needed threads. It should engage the sliding nut fully, but it should not touch the bottom of the Tee slot.

The Tee slot is longer than you can use. As you move the crank pin away from the center of the bull wheel, the stroke is made longer.

The position of the lower pivot will greatly effect the length of the stroke. Fundamental laws of leverage are involved here; the lower pivot is the fulcrum of the lever.

The object is two fold: To position the lower pivot as high as possible, and still have clearance between the top of the yoke and the ram slide, and to center the pivot fore and aft so the yoke will have clearance on both ends of the stroke.

Slip a short piece of 1/4" stock in the ram slide channel, so you can see that the yoke clears by 1/8" as it passes the highest point in its travel.

Lock the crank pin about 1 1/4" from the center of the bull wheel, and totate the pulley by hand so you can measure the clearance at each end of the stroke.

61

The crank is to rotate clockwise as you face the left side of the column, and clearance is measured between the yoke and the front column casting, and between the yoke and the top rear spreader.

If the clearance is not equal on both ends of the stroke, move the lower pivot center forward or back a small amount, and test again.

When the lower pivot is properly positioned, re-check the top clearance.

As you adjust the position of the lower pivot center, be sure that the yoke remains parallel to the column side.

When the clearance is equal, both fore and aft, and you have clearance at the high point of the stroke, re-adjust the crank pin to give a longer stroke. Rotate one full turn after each adjustment, and find the crank pin position for the longest possible stroke. When you have the stroke long enough to leave just 1/8" clearance at each end, rotate the wheel until the crank pin is at its highest point, and the Tee slot is parallel to the center of the yoke. Make a dot on the yoke leg with a center punch, to mark the top edge of the slide block. This will mark the maximum stroke setting. Later, we'll calibrate the yoke so you can pre-set the stroke length a definite amount.

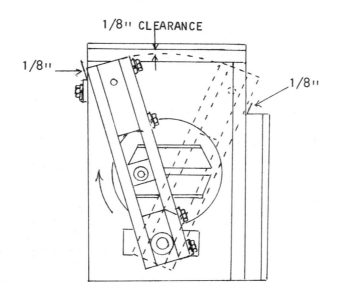

1/8" CLEARANCE

1/8"

1/8"

The yoke will actually travel through the " U " shaped opening at the top of the front casting, but that won't be any problem.

The final check is to make sure the yoke is parallel to the column sides.

Double check everything, then step drill 1/4" holes in each side, through the column side, and the flanges on the bosses.

Enlarge the holes in the side castings to 5/16", and tap the holes in the flanges for 5/16"-18 threads.

Be careful not to enter the flanges with the 5/16" drill, and use the holes in the side casting to guide the tap.

Install four 5/16"-18 X 1" cap screws with lock washers.

LONGEST STROKE

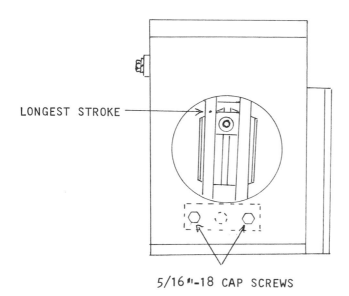

5/16"-18 CAP SCREWS

THE FAST RETURN STROKE

Because the lower pivot is closer to the center of the crank than the upper pivot, the return stroke is faster than the forward stroke. If you reverse the rotation of the crank the forward stroke will be faster.

In early years, the makers of shapers had a lot to say

about their " Fast Return Stroke " shapers. Since it must make both a forward and return stroke to each cycle, I fail to see how a fast return stroke would have any effect on the production capacity of the machine, even though the " Hot Air " salesmen had a lot to say about it.

It's a fascinating action to study though, and it makes it a lot more fun to demonstrate the machine. See if your friends can figure out why it does it.

THE RAM CLAMP

Two 7" lengths of 5/8" key stock make up the main body of the clamp. The rear spreader and the guide are scraps of 1/2" key stock, and the front spreader is two 2" lengths of 1/4" X 1" cold rolled steel. The assembly is fastened with three 1/4" X 2" flat head machine screws.

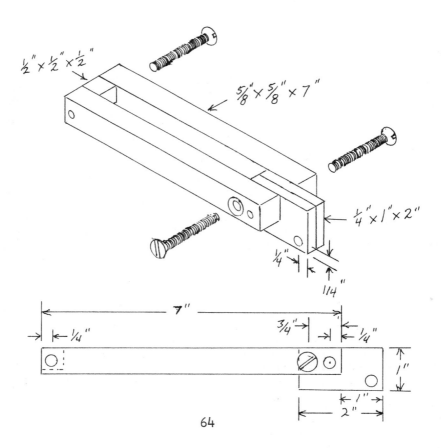

Notice that the second screw in the front end is made to enter from the left side.

Like all stacked assemblies, clamp them together, step drill the holes to tap size, enlarge through holes to bolt size, and tap through the through holes. Again, tap very carefully.

The 1/4" link hole should be as straight as possible, so the link won't bind.

The top of the assembly needs to be all flush.

THE LINK

Two lengths of 1/4" X 1/2" cold rolled steel. The ends are rounded, and the holes are step drilled to 13/64" through both members. The holes are enlarged to 1/4" in the left half, and tapped for 1/4"-20 threads in the right half.

The pivots are 1/4"-20 cap screws, with a jamb nut on the right hand plate to lock them in place.

If you can't find any cap screws that have an unthreaded portion 3/4" long, make them from longer cap screws.

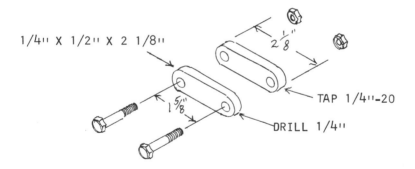

1/4" X 1/2" X 2 1/8"

2 1/8"

TAP 1/4"-20

1 5/8"

DRILL 1/4"

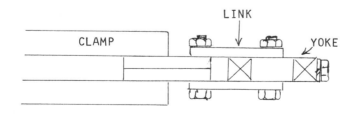

LINK

CLAMP

YOKE

Adjust the link pivot screws for a free fit with minimum side play, and tighten the jamb nuts against the right hand side.

THE CLAMP GUIDE AND BOLT

The guide must be located on the bottom of the ram by a trial assembly. It guides the ram clamp so that it will remain aligned with the yoke as it is moved through its range of adjustment.

A 1 1/4" length of 1/2" key stock is filed so it will be a free sliding fit between the rails of the clamp.

A 1/8" pilot hole is drilled 3/8" from each end.

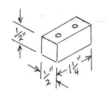

Install the ram in its slide channel, and raise the clamp against the bottom of the ram. Hold it securely with a " C " clamp, and rotate the crank through one stroke to make sure all is tracking correctly.

Scribe a sharp clear mark inside the rails of the clamp, and remove the ram to install the guide.

Position the guide in the marks, 2" from the tail end of the ram, and clamp it securely.

Step drill a 13/64" hole through the front pilot hole, and into the ram.

Enlarge the hole in the guide to 1/4", and tap the hole in the ram for 1/4"-20 threads.

Countersink the hole in the guide, and install a 1/4"-20 X 3/4" flat head machine screw.

Make sure the guide has not slipped, and step drill a 5/16" hole through the rear pilot hole, and about 1" into the ram.

Enlarge the hole in the guide to 3/8", and tap the ram for 3/8"-16 threads, about 3/4" deep.

The clamp bolt is a 3/8"-16 X 1 1/2" cap screw, and the clamp plate is a 1 3/4" length of 1/4" X 1" cold rolled steel with a 3/8" hole drilled in the center.

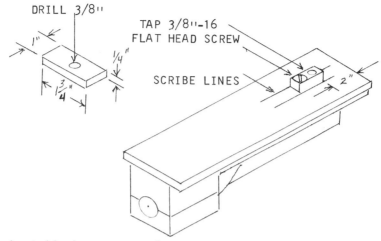

DRILL 3/8"

TAP 3/8"-16
FLAT HEAD SCREW

SCRIBE LINES

1"

1/4"

3/4"

2"

Re-install the ram, and raise the clamp to engage the
guide. Install the clamp bolt and plate.

The guide has been positioned to permit the ram to begin
its stroke farther back than ordinary design would dictate.

This is so the machine will be able to plane its own work
table.

At a later time, you may want to re-position the guide
so you won't inadvertently start the stroke too far back.

CALIBRATE THE YOKE

The longest possible stroke has been marked on the leg
of the yoke to prevent exceeding the capacity of the shaper.

Now you can calibrate the yoke in 1/2" increments so you
can pre-set the stroke length accurately.

The marks are easiest to understand when they are read
at the highest point of the slide block, so plan a vertical
row of dots below the limit dot.

Make one dot for 1", one dot for 1 1/2", two dots for
2", one dot for 2 1/2", etc..

Just adjust the crank pin, and measure the length of the
stroke until you have the right divisions.

A bit tedious now, but it will save a great deal of time
later on when you are using the machine.

67

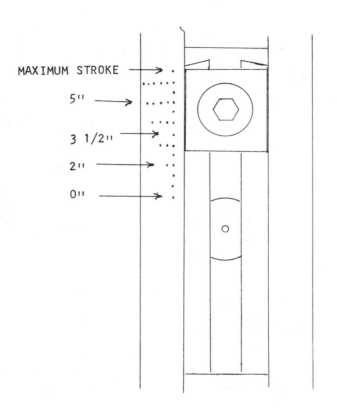

MAXIMUM STROKE ⟶

5" ⟶

3 1/2" ⟶

2" ⟶

0" ⟶

THE FEED CRANK

Another pattern that has draft above and below the parting line. It will be easiest to mold as a split pattern.

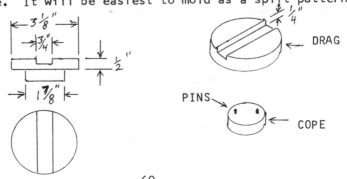

The parting is at the junction of the two discs, and it requires a small fillet at the junction.

When the discs are prepared to size, align them with the pins, like all split patterns, and assemble them with waxed paper in the parting, so you can wipe the fillet on the cope half of the pattern.

When you ram up the drag half, be sure to press sand in the 1/4" X 3/4" channel.

Roll over and set the cope half in place. Make sure the halves will separate easily when you open the mold.

Set a 1" riser in the center of the hub, and a 1" sprue about 1" away from the side.

Ram up and vent the cope, remove the pins, rub in a bottom board, and open the mold.

Lay the cope down to swab, rap, and remove the pattern.

Swab the sprue opening to form a fillet at the junction.

Swab, rap, and remove the drag half of the pattern, cut the gate, and close up the mold for pouring.

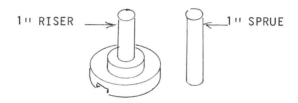

1" RISER   1" SPRUE

MACHINING THE FEED CRANK

Cut off the sprue and riser, and punch the center of the hub to locate the bore.

The casting can be drilled and tapped, or you can mount it on the face plate with clamps.

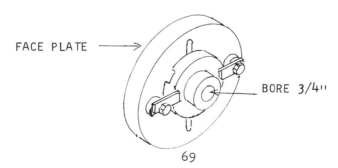

FACE PLATE   BORE 3/4"

69

Bring up the tail stock center to position the casting on the face plate, and drill a 3/8" starting hole with the tail stock chuck.

Enlarge the bore to exactly 3/4" with the boring bar.

Drill and tap a 1/4"-20 set screw hole in the hub, and lock the casting on a 3/4" arbor.

Mount the arbor between centers, and face off both ends. True up the diameter of the plate, and reduce the diameter of the hub to 1 3/4".

Two 1/8" X 3/4" steel rails are fastened to the plate with four #10-24 flat head machine screws to form a Tee slot for the crank pin sliding nut.

The sliding nut is 1/8" X 3/4" X 3/4" steel, tapped for 1/4"-20 threads.

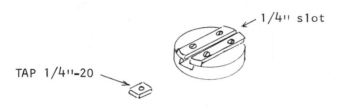

1/4" slot

TAP 1/4"-20

INSTALL THE FEED CRANK PLATE

Remove the temporary set screw collar from the crank shaft, and install the feed crank plate in its place.

Cut off any excess amount of crank shaft, so the feed crank hub will rest against the bearing flange with no end play in the crank shaft.

Don't lock the feed crank tightly to the shaft at this time. Its position, in relation to the ram crank, will have to be determined when the feed ratchet is installed.

ON TO BIGGER AND BETTER THINGS

This completes the main body and mechanism of the shaper. The remaining work involves considerable detail, but none of it is unusually difficult.

Many trained machinists will declare that this project is beyond the scope of even a very well equipped shop. By now you have seen that it is well within your ability, and you have gained skill and knowledge for even greater things.

CHAPTER IV

THE TOOL HEAD AND DOWN FEED

This is a really exciting group of excercises in molding, casting, and machining.

A number of the features that are included didn't seem possible when I began to design the project. I continue to be amazed at the range and capacity of work that can be done with simple methods and equipment.

THE ROTATING HEAD

In order to form dovetail slides, and other work that requires an angular down feed, the head is made to rotate on an arbor. The arbor fits the bore of the ram, and it's locked in any position by the set screw.

The base of the rotating head will bear against a cast disc which is fastened to the end of the ram. By graduating the disc, it's possible to accurately set the angle of the down feed.

GRADUATING THE DISC

Without a dividing fixture, it would be very near impossible to accurately graduate the disc.

The only practical method is to use something that is already equally divided, and apply it to our purpose.

This path of reasoning may seem elusive at first, but it will prove valid as you follow it through.

To begin with, there are 360 degrees in any circle. If we can divide a circle into 360 equal graduations, each one will represent one degree.

Let's select an ordinary tape measure and see how it is possible. A convenient division is sixteenths of an inch. We need 360 of them so, 360 divided by 16 is 22 1/2". The circumfrence divided by 3.1416 equals the diameter. In this case it turns out to be slightly over 7".

7" is too large for the job at hand, but 1/32" divisions work out to a diameter slightly over 3 1/2". Just right for this job.

It's a simple matter to mount an oversize disc, and very carefully reduce its diameter until the circumfrence is exactly 11 1/4", which is 360 1/32" divisions.

The head can be rotated through a full circle, but only half of the rotation has any practical use. We only need to graduate the top half of the disc.

A 6" flexible steel rule, with 1/32 divisions, can be fastened to the disc with a couple of small machine screws, and an adjustable indicating pointer is fastened to the head to make a very accurate protractor.

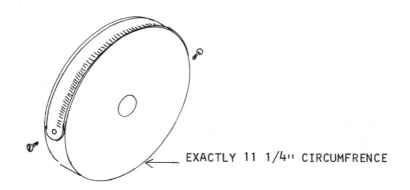

EXACTLY 11 1/4" CIRCUMFRENCE

The three inch mark is the center, or zero point on the protractor. When the shaper is finished, you can make a series of test cuts until the down feed cuts at exactly 90 degrees to the work table. Then you set the pointer to the three inch mark to zero the scale.

The most common angles are 15, 30, 45, and 60 degrees. You will quickly learn to use the scale as a protractor.

GRADUATING FEED COLLARS

The same principle is applied to the graduated collar on the down feed screw.

As we will be using a 7/16"-20 screw, each full turn of the screw will advance the cutter 1/20", or .050", so we need 50 divisions to represent .001" graduations.

There are 50/16" in 3 1/8", and that is very near to the circumfrence of a 1" collar.

Just reduce the diameter of the 1" collar until 3 1/8" exactly reaches around it, and you have 50 graduations.

You can use 1/8" divisions to make a larger collar, or you can use a metric scale for very fine divisions.

72

CAST ON STEEL ARBORS

The design is similar to the compound casting for the lathe, but it's larger and heavier.

Both the protractor base and the rotating head are cast on a steel arbor, which provides the means to machine them.

Two 3/4" X 6" arbors are prepared from oversize steel stock. They must be made to exactly 3/4" diameter, with 60 degree centers in each end.

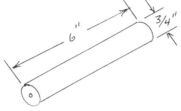

THE PATTERNS

A three piece pattern will mold both the protractor and the rotating head.

The core print, which serves both castings, is best made by machining a 1" hardwood dowel rod between wood turning centers on the lathe. The 1/2" shoulder needs to be concentric with the 3/4" main body. It becomes a part of the pattern when it is slipped into the 1/2" hole in the center of the disc. It leaves a 2 3/4" deep print in the mold, into which the steel mandrel is set as a core.

The disc can be made by mounting a 3 3/4" disc of 1/2" plywood on the threaded adapter, and machining it to 3 5/8" diameter on the lathe. It needs a slight amount of draft, while the core print is made without any draft.

The pad that supports the down feed slide is a simple rectangle with a tunnel for the down feed screw. The tunnel is 5/8" X 2" at the surface, and it tapers 5/8" deep to a rounded bottom, so it will release the green sand core easily.

There are two alignment pins in the base of the pad, to align it with the disc. The overhanging end of the pad must be no more than 3 1/2" from the center of the disc, so its size will be within the capacity of the lathe. The alignment pins must fit holes in the disc freely, so the halves will separate easily when the mold is opened.

Assemble the pad and disc with a piece of waxed paper between them, and wipe a small fillet to the three sides of the pad that join the disc. Both surfaces of the disc are to remain flat.

The parting line is at the junction of the disc and pad, and you need slight draft both above and below the parting.

Align all members carefully as you assemble them.

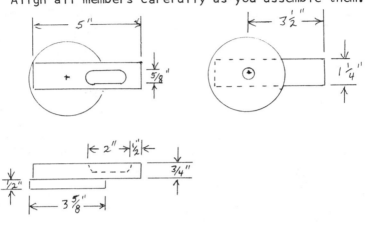

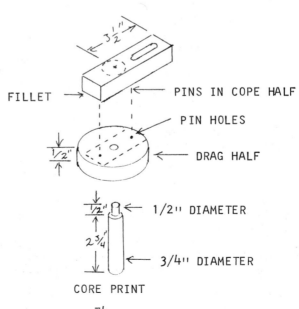

FILLET →   ← PINS IN COPE HALF

← PIN HOLES

← DRAG HALF

← 1/2" DIAMETER

← 3/4" DIAMETER

CORE PRINT

MOLDING THE ROTATING HEAD

The only tricky part of this mold is that the green sand core that forms the tunnel will be suspended from the cope.

The tunnel in the pattern must have extra draft, and it must be very smooth, with no undercuts in the surface.

The drag half of the pattern is layed, pin holes down, on the molding board, and the core print is slipped into the 1/2" hole in the center.

Ram up and vent the drag, rub in the bottom board, and roll over.

Set the cope half in place, and make sure it will separate easily.

Set a 1 1/4" riser in the center of the pattern, and set a 1" sprue pin about 1 1/2" away from the tail of the pad.

Press sand into the tunnel with your fingers, and push two four penny nails, with their ends bent, into the sand. The heads of the nails will be bedded in the tunnel, and the bent tails will anchor the core to the cope sand.

Ram up the remainder of the cope carefully, and be sure you don't strike the nails.

Also, be careful not to strike the riser so the cope half of the pattern won't be jarred loose before you open the mold.

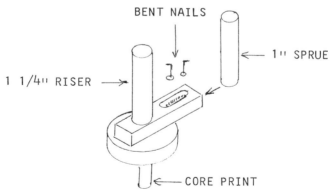

BENT NAILS

1" SPRUE

1 1/4" RISER

CORE PRINT

Remove the sprue pin and riser, and rub in a bottom board before you open the mold.

Lift the cope very carefully, and lay it down to swab, rap, and remove the pattern.

Swab, rap, and remove the drag half of the pattern, and

clean up both cavities before you set the core.

Fill the center holes of the steel core with graphite before you set it in the mold. The core will stand above the parting line partially, and its end will be in the riser when the mold is closed up. It will be surrounded by aluminum, but it will break off easily to expose the end of the mandrel. The graphite will prevent the center from filling with aluminum.

Cut the gate, blow out the sprue and riser, and close up the mold. Pour rapidly, as with all molds.

MACHINING THE ROTATING HEAD

Break off the riser to expose the mandrel center, and cut off the gate.

Mount the mandrel between centers, and rotate it by hand to make sure the tail of the pad will clear the bed. If it does not, trim its length until it does.

Face off the top of the pad and the base of the disc. The disc is reduced to 3/8" thick, but remove only enough to clean up the top of the pad.

True up the diameter of the disc, and reduce it to about 3 3/8".

Clean up the mandrel with emery cloth, and slip it into the end bore of the ram. If the disc is too large to clear the slide clamps, return it to the lathe and trim it until it will clear by about 1/32" on both sides.

At this point, the mandrel will seem too long, because the protractor disc is not yet in place. That's the next order of work, and it must be done before the front end of the mandrel is cut away.

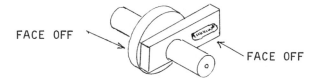

FACE OFF

FACE OFF

The disc on the head carries the indicating pointer for the protractor, its finished diameter is not critical.

THE PROTRACTOR DISC

This uses the same disc pattern as the rotating head, but the pad is eliminated.

The molding and coreing procedure is the same, and it is prepared for machining in the same way.

MACHINING THE PROTRACTOR

Face off both sides, and reduce the thickness to about 7/16".

I stole a plastic tape measure from my wifes sewing box, and cut an 11 1/4" length from it.

I reduced the diameter of the disc until the tape would just reach around it. That gives a circumfrence of 360/32".

I've been forced to replace the tape measure, but it was only 69¢ at the dime store, and I've been forgiven.

When the disc is finished to size, grind a bevel on the short end of the mandrel, and put a couple of drops of oil on each side of the casting.

Support the casting close to the mandrel, and strike a couple of sharp blows on the long end to break it loose.

Turn it over and drive the mandrel out of the disc with a drift punch.

INSTALL THE PROTRACTOR ON THE RAM

Lightly scrape the bore of the protractor until it is an easy fit on a 3/4" mandrel.

Use the mandrel to center the protractor on the front of the ram, and step drill two 13/64" holes through the disc, and about 1" into the front of the ram.

Enlarge the holes in the disc to 1/4", and countersink them for flat head machine screws.

Tap the holes in the ram for 1/4"-20 threads, and install the protractor with two 1/4"-20 flat head machine screws.

Two notches are cut at the bottom of the disc so it will clear the ram slide clamps. Mark them from a trial assembly, and cut them with a hack saw.

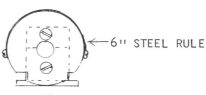 ←—6" STEEL RULE

MACHINE THE HEAD ARBOR

The 3/8" set screw in the ram cap will lock the arbor at its setting. Because the screw will soon mar the arbor, a groove must be machined in it so the burrs will not jamb up the movement.

Mark the arbor through the set screw hole, and mount it between centers to machine a groove about 1/16" deep at the point that the screw contacts it.

When the arbor and tool head fit properly, re-mount it in the lathe and begin to cut the front end of the arbor off flush with the top surface of the pad. Use a hack saw to cut a deep groove, but don't attempt to finish the cut in the lathe. Cut off work is never done between centers. The groove merely serves as a guide so you can finish the job in a vise. BE VERY CAREFUL OF THE SPINNING WORK AS YOU MAKE THE SAW CUT IN THE LATHE.

INSTALL THE PROTRACTOR POINTER

The disc of the head is smaller than the protractor, so the pointer must be thick enough to make up the difference in diameters.

A scrap of aluminum about 3/8" wide and 1" long will do.

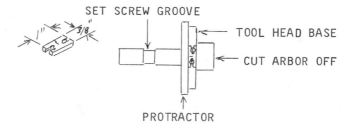

Cut slots in each end of the pointer plate, so it will be adjustable, scribe a sharp mark in the center, and fasten it to the tool head base with two small machine screws.

For an initial setting, you can lock the tool head so that the side of the pad is parallel to the vertical slide, but you will need to make a fine adjustment when the shaper is complete.

THE DOWN FEED

Very similar to the compound slide on the lathe, but it is easy to mold as a one piece pattern, because it has no raised portion on the top surface.

Like all of the hand fit slides, it has a recess between the wear pads to reduce the labor in scraping.

The exploded view shows the elements of the pattern. You can use layers of cardboard to build up the wear pads if you don't have an easy way to cut wood to 1/16" thick.

There is draft in only one direction, since the parting line is at the broad flat surface.

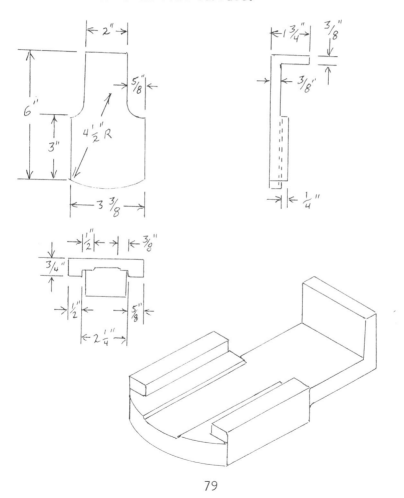

Round all outside corners, except at the parting plane, and wipe small fillets on all inside corners.

Only minimum draft is required, but make sure that the screw bearing support is exactly at right angles to the tail of the slide, or it will be difficult to draw from the mold.

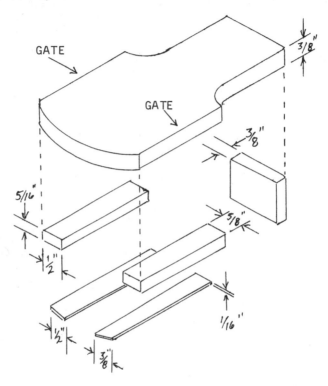

MOLDING THE DOWN SLIDE CASTING

No special tricks, just lay the pattern on the molding board, and ram up and vent the drag.

Roll over, and set a 1" sprue pin on each side of the pattern, about 1" away.

Ram up and vent the cope, open the mold, and draw the pattern.

Cut full size gates to each sprue pin, clean up and close the mold.

FITTING THE SLIDE

A slab of 1/4" X 2" X 5" cold rolled steel is fastened to the rotating head to form the slide ways, and this will be the test standard for scraping the slide.

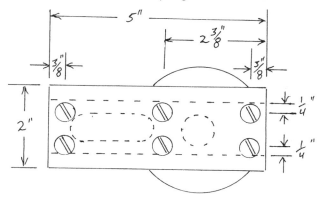

The pad on the rotating head should be scraped true flat before the steel ways are installed.

Center the ways on the pad so it overhangs equally on each side, and locate the centers of the screws carefully, to clear both ends of the tunnel, and to be 1/4" inside the edge of the pad.

Step drill as usual, tap the pad, and install the ways with six 1/4"-20 flat head machine screws

Countersink the heads just enough to be below the surface.

FIT THE SLIDE TO THE WAYS

Prepare the box slide for scraping by filing off the sand texture, and cut diagonal grooves in the corner to separate the pads.

You can cut down the drastic high spots by laying a sheet of emery cloth over the ways and using it as a sanding block.

Then blue the slide, and use the ways as a test standard as you hand scrape the slide to a good fit.

The gib goes in the left side, so only the right side needs to bear on the ways.

81

THE SLIDE CLAMPS

Two 3" lengths of 1/4" X 1" cold rolled steel. They are fastened with six 5/16"-18 X 1" grade 5 cap screws with lock washers.

Like the ram slide, the clamp pads are filed parallel to the ways, and brought several thousandths of an inch lower so shims can be installed under the clamps.

There is a 1/8" X 1/4" gib in the left side, and two gib screws, which are #10-24 machine screws with jamb nuts. The screws are ground to a point, and they seat in dimples in the gib.

Clamp together and step drill. Same procedure as the ram slide clamps.

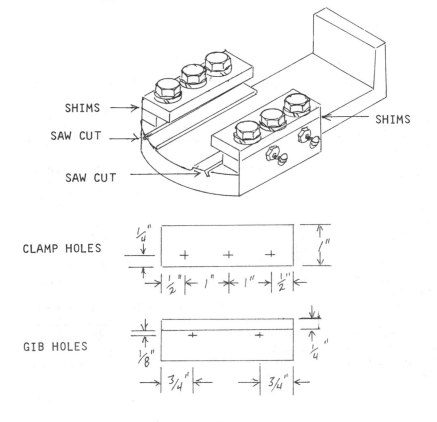

SHIMS

SHIMS

SAW CUT

SAW CUT

CLAMP HOLES

$\frac{1}{4}$"

1"

$\frac{1}{2}$" | 1" | 1" | $\frac{1}{2}$"

GIB HOLES

$\frac{1}{8}$"

$\frac{1}{4}$"

$\frac{3}{4}$" | $\frac{3}{4}$"

THE DOWN FEED SCREW

Prepare 60 degree centers in a 5 1/4" length of 3/4" steel stock, and mount it between centers to machine the screw blank.

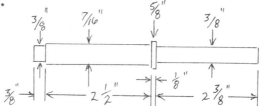

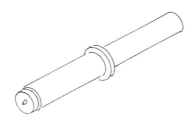

As your home made lathe lacks screw cutting capability at this time, the 7/16"-20 threads must be cut with a die.

Use the type of die stock that has an adjustable centering guide, in order to make the threads as true as possible.

The short 3/8" stub will serve to start the threads concentrically. It is cut off when the screw is finished.

It will work best if you machine the 7/16" end of the blank first, cut the threads with the work held in a vise, and return it to the lathe to finish the other end.

The threads are easier to cut if the blank is slightly undersize.

When you near the shoulder, back the die off, and turn it over to cut threads right up to the shoulder.

Use plenty of oil when you cut the threads, and back off from time to time to clear the chips.

Run a nut onto the screw before you cut off the stub, and grind or file a bevel on the end of the screw.

INSTALL THE DOWN FEED SCREW

This job must be done with extra care, because there is not much clearance in the screw tunnel. It's done exactly the same way it was done on the lathe.

Locate the center of the tunnel, and punch the center 1/4" below the surface of the slide pad.

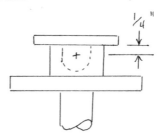

Scribe a line across the face of the screw support that is parallel to the back surface of the clamps.

PARALLEL TO CLAMP SURFACE

Measure the distance from the non gib side of the ways to the center of the screw tunnel, and transfer the dimension to the screw support.

This will be very near to 1", but an error as slight as 1/16" may mean that the screw will run into the side of the tunnel before it bottoms.

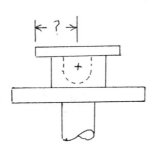

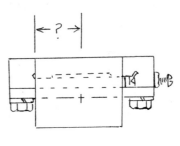

84

If the last center measurement is transferred by laying a straight edge along side the ways, with the slide assembled and the gib screws snug, the centers should be nearly perfectly aligned.

Punch the center, and drill a 1/8" pilot hole through the screw support.

With the slide assembled on the ways, and the gib screws snug, use the hole in the screw support to guide the bit as you drill into the center of the tunnel.

Drill as deep as the 1/8" bit will reach, and repeat with a 3/16" bit. Continue to step drill through both members until both holes are 3/8".

Using the hole in the screw support to guide the shank of the tap, thread the hole in the head for 7/16"-20 threads.

An amount of error can be tolerated in these operations, but use extra care.

If the screw runs into the tunnel in spite of your care, you can widen it for clearance.

If the screw is eccentric, you can ream the hole in the screw support.

THE FEED HANDLES

I used the 3" ball handles that are described in " Charcoal Foundry " and " Metal Lathe ".

You already know how to make them, so there's no need to go into that again.

The shaper uses three of them, and they all have a 3/8" bore. They are locked with 1/4"-28 socket head set screws.

GRADUATED FEED COLLARS

As the down feed screw is a 20 pitch, the feed will be advanced 1/20", or .050" for each full turn of the screw.

50 divisions on the collar are required for .001" graduations.

A standard 3/8" set screw collar is 3/4" 0. D., and that turns out to be very near 50 MM in circumfrence. You can reduce the diameter of the collar until a 50 MM length of tape measure will fit around it perfectly, and cement the tape to it for a simple collar.

A sleeve with an 0. D. of slightly less than 1" will be 3 1/8", or 50/16" in circumfrence. You can prepare it to a friction fit over the collar, and cement a scale with the 1/16" divisions to the sleeve.

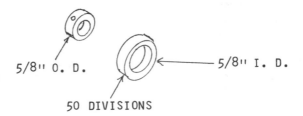

5/8" O. D.　　　　　　　　　　　　　　　← —— 5/8" I. D.

50 DIVISIONS

    You must fit the sleeve so it can be removed for access to the set screw in the smaller collar.

    Other combinations are possible, or you can add a collar to your ball handle pattern, and cast them together.

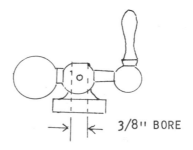

3/8" BORE

    The ball handle can be mounted between centers on an arbor, to machine the collar on the lathe.

THE FEED DIAL POINTER

    You can file the screw support so the dial will bear evenly all around, and scribe a mark on the support, but it will work better if you prepare a boss to screw to the support.

3/8" BORE ⤴　　　　　　　　← DIAMETER OF COLLAR

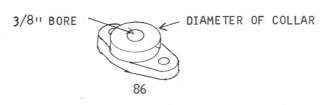

86

The casting can be molded with a rectangle and a riser just like the yoke pivot support bosses. Bore a 3/8" hole, and machine it on an arbor.

The screw support will have to be filed so the collar will bear evenly all around.

A 3/8" S. A. E. washer is slipped on the shank of the screw, to bear against the screw support.

File a flat on the screw shank, for the collar set screw to bear against, and cut away the excess shaft length when all is fit together.

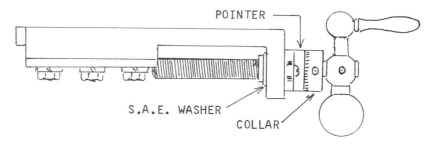

POINTER

S.A.E. WASHER

COLLAR

THE CLAPPER BOX AND BLOCK

Probably the most demanding job in the project, the clapper must fit the box without friction, and without play.

There will be considerable filing and scraping to the box and block, but most of the rest of the work can be done on the lathe.

THE CASTINGS

The drawings give all the necessary details, and there are no special tricks to casting them.

The box is molded just like the down feed slide, and it is gated in the same way, on both sides.

The gates must be full size to avoid a shrink cavity at the junction.

The block can be molded with a 1 1/4" riser in the center, and it can be poured through the riser. It's made oversize, to provide plenty of stock for machining.

Provide minimum draft on both patterns, especially the channel between the box members, to avoid unnecessary work in finishing them up.

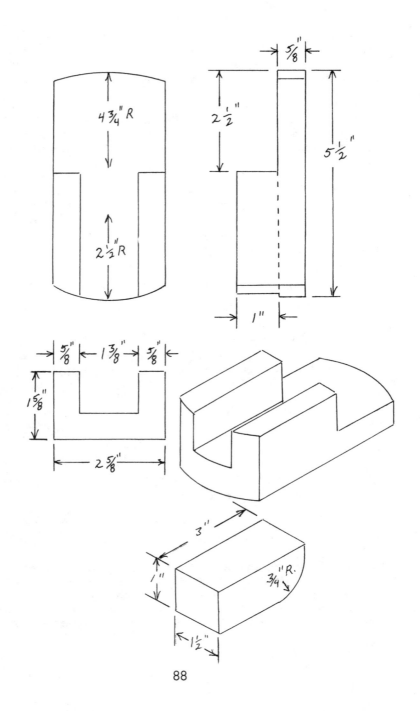

Of course the outside corners must be rounded, except at the parting line, and inside corners need a small fillet.

MACHINING THE BOX

The initial work can be done quickly on the lathe. Drill two 1/4" holes in the base, and tap them for 5/16"-18 threads.
These will be used to mount the box on the face plate, and the hole between the uprights will be used to mount the box on the down slide. The upper hole will become part of the elongated hole for the top bolt.

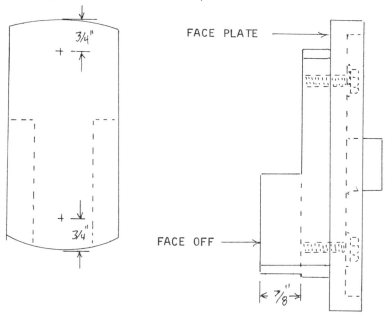

File any roughness from the base of the casting before you mount it on the face plate with two 5/16"-19 cap screws and flat washers.
Face off the uprights, and reduce the height to 7/8".
Invert the casting on the face plate, and fasten it in the same bolt holes with longer bolts. Face off the base of the casting, and reduce its thickness to 1/2"
You can re-mount the base, and face off the upper part of the tongue. It must be mounted off center with one bolt, and be careful of the uprights.

A 3/8" elongated hole is needed at the top of the tongue, and it's layed out on a 4" radius from the lower hole.

Just punch and step drill a row of 3/8" holes, and finish with a coarse file.

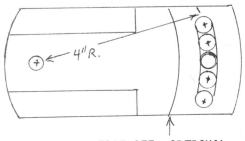

FACE OFF, OPTIONAL

Make diagonal saw cuts in the inside corners of the uprights, and do some preliminary filing to clean up the channel, but leave the major finishing of the box until you have made a gauge to file the inside square and parallel.

After preliminary filing, estimate the finished width of the inside of the box that will require the least work.

Make a gauge of heavy sheet metal in the shape of a large " T ", which will be a pair of try squares and a depth gauge all in one.

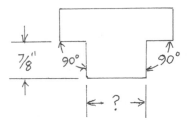

Be especially sure that the sides are parallel, and the corners are true square. Make the estimated width an even decimal, so the dimension will be easy to transfer.

Use the gauge to finish the inside of the box, and take plenty of time to do the work well.

The finished width of the box will be the width of the block, minus .001". The final fit will have to be done by hand scraping.

90

MACHINE THE CLAPPER BLOCK

The face plate angle clamps will do the job here.   If you are lucky enough to have a lathe with a four jaw chuck, so much the better.

The portion with the rounded corner is the bottom, and it should be faced off just enough to clean it up.

Face off the opposite side, and reduce the thickness to 7/8". Take special care to see that the front and back are parallel when you face off the second side. If there is an error in the clamp steps, now is the time to catch it.

Face off the third side of the block, just enough to clean it up, and check to see that the corners are exactly 90 degrees.

Face off the fourth side, parallel to the third side, and reduce the width to a snug fit in the box.

The remainder of the fitting is done by scraping, and you want it to fit the channel without friction, and with no perceptible side play.

BORE THE TOOL POST HOLE

Mark the location of the hole and punch it. Mount the block in the face plate clamps, and bring up the tail stock center to center it on the face plate.

Using the tail stock chuck, step drill a 3/8" starting hole through the block.

Enlarge the hole to 11/16" with the boring bar, and bore the 7/8" shoulder 3/16" deep.

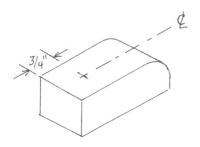

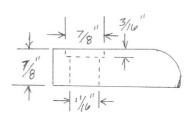

THE TOOL POST

An easy job, done between centers on the lathe.   Just prepare centers in a 6" length of 1" diameter steel, mount between centers, and reduce the diameter to 7/8" for about 3" from the tail stock end.

Face off the tail stock end, and reduce the diameter to 11/16" for 2 3/8".   Bevel the tail stock end.

Score a groove at the 2 1/2" point, and cut about half way through with a hack saw.   BE CAREFUL OF THE SPINNING CLAMP DOG, AND REMEMBER THAT YOU CAN'T CUT COMPLETELY THROUGH THE WORK.

Before you cut the post from the stock, drill a row of three 9/32" holes for the tool slot.

File away the waste, and finish the slot to 5/16" wide, and 1" long.

Step drill the center hole through to the slot, and enlarge it to 1/4".   Tap the hole for 5/16"-18 threads.

Now you can cut away the waste, and the tool post is complete.

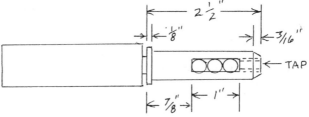

A 5/8" wrought washer will fit the post, and a set screw will lock a 5/16" square lathe tool bit in the slot.

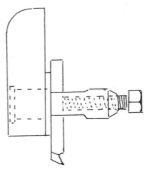

THE CLAPPER HINGE

The clapper block pivots at a point 3/8" from the top, and 3/8" from the front of the block.

The pivot pin must be at exact right angles to the inside of the box, or the block will bind when it lifts on the return stroke.

In the absence of a very good drill press, and a perfect vise or holding fixture, the job will have to be done on the lathe face plate with the tail stock drill chuck.

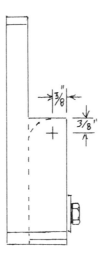

The block can be held in the box with a 5/16"-18 X 1 1/2" cap screw and flat washer

The first step is to face off the outside of the box so that it is parallel to the insede. This is probably best done before you bolt the block in the box.

Mount the box in the face plate clamps, and measure from the inside of the box to the face plate, to set up for the first side.

When you are certain that the box is parallel to the face plate, face off the exposed side. Check your work as you go, to make sure it's right.

When you turn it over to face off the other side, the steps in the clamp should register the work parallel.

When both sides are parallel to the inside of the box, bolt the block in place, punch the pivot center, and align the work to the lathe center with the tail stock center.

The pivot pin will be a 2 1/2" length of 3/16" drill rod, and it would be best to step drill the pivot hole under size, and finish with a reamer, but you can do it with a drill for a slightly less accurate clapper.

You can also make an oversize pin in the lathe, to fit the drilled hole. The object is to have a free swing with no perceptible play.

Take extra care when you set up to drill the pivot hole. Make sure that the inside of the box is parallel, and also that the base of the box is at right angles to the face plate.

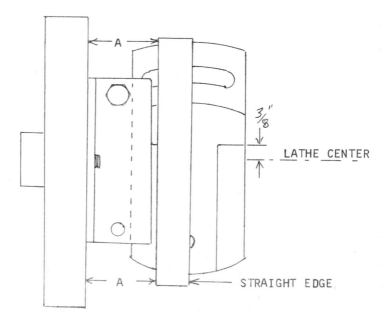

A = SAME AT BOTH ENDS

When all is true, bolt the block in the box, and step drill the pivot hole.

The pin can be locked with a small set screw in one of the uprights, or you can stake the holes with a punch.

FINISH THE DOWN FEED SLIDE

The front of the down feed slide is faced off parallel to the clamps before the clapper box is mounted.

Remove diagonally opposite clamp screws, and bolt the slide to the face plate, with spacers to clear the screw support. A pair of standard 3/8" set screw collars will be the right amount of space.

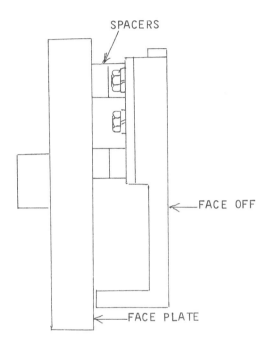

SPACERS

← FACE OFF

← FACE PLATE

The surface does not need to be entirely perfect, don't remove excessive material and make it too thin.

The lower pivot is a 5/16" flat head machine screw, and it is locked with a short set screw in the clapper box that bears against the end of the machine screw.

The top hole is tapped for 3/8"-16 threads, and a cap screw and flat washer lock on the elongated hole to set the angle of the box.

Step drill both holes to 5/16" at the locations shown in the drawing.

Countersink the lower hole from the back side, so the head of the machine screw will be even with the surface.

A 3/4" long screw should reach about half way through the base of the clapper box, and the short set screw is to lock it in place when you have adjusted it to a snug fit.

It may be necessary to trim the length of the screw.

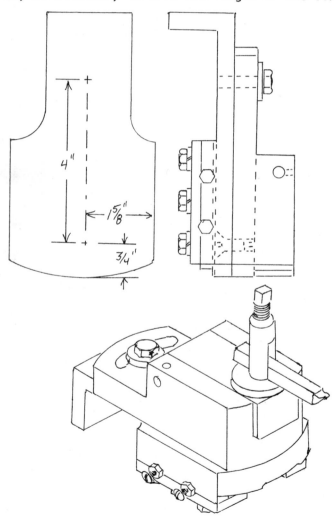

COMPLETE THE DOWN FEED

All of the components for the tool head are finished. It remains only to assemble them and install the head on the ram.

Install the slide on the down feed ways, and adjust the gib so there is a slight drag in the feed.

Adjust the feed screw collars so there is no end play in the motion, and file a flat on the shank of the screw for the handle and collar set screws to seat against.

Install the completed tool head on the ram. The base of the tool head should seat squarely against the protractor disc, and there should be a slight drag when you loosen the set screw to rotate the head.

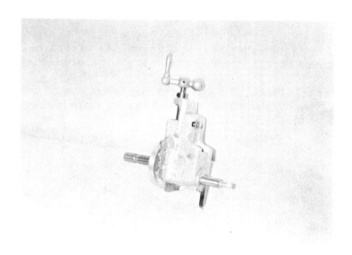

CHAPTER V

THE WORK TABLE AND SLIDE

The cross slide support casting is made to ride the vertical slide ways, and it is raised and lowered by a screw.

This is a rather heavy casting, and it has a box slide that is fitted to the vertical ways, and a pair of pads to support the cross slide ways.

The cross slide ways should be at exact right angles to the vertical ways, so you should take extra care as you make the patterns.

There is considerable detail in this phase of the project, but they are all relatively simple jobs.

THE CROSS SLIDE SUPPORT CASTING

Another figure that can best be molded with a split pattern, because it has draft both above and below the parting line.

At first glance, the pattern may seem complicated, but it is easy to make when you view it as simple rectangles to be joined together.

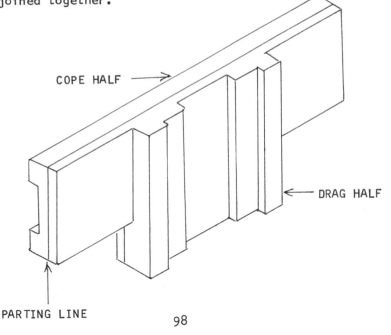

COPE HALF ⟶

DRAG HALF ⟵

↑
PARTING LINE

98

THE DRAG HALF PATTERN

The foundation of each half is a simple " T " shape, and both halves are identical, so make two of them.

The important thing is to use stock that is flat, so the halves will mate well together. 3/8" exterior glued plywood is OK, or you can rip 3/8" stock from pine.

Square all corners accurately, and make sure the thickness is uniform in each piece.

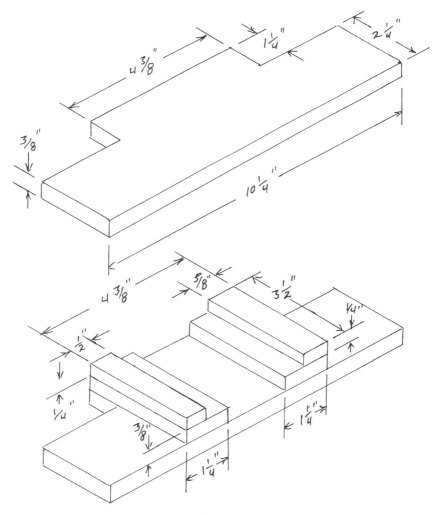

Notice that the left clamp pad is not as wide as the right side, because there will be a gib in the left side.

The space between the clamp pads is 3 3/8", but it will shrink along with the entire length, and the casting will be the right width for the 3" vertical ways.

Assemble the parts with glue and brads, and square them up carefully before you fasten permanently.

THE COPE HALF PATTERN

Made on the same basic foundation as the drag half, but don't forget, they fit together as opposing members.

The small block is fit between the rails to make a base for a large riser, because extra metal must be available as the casting solidifies, or there will be a bad shrink cavity in the center.

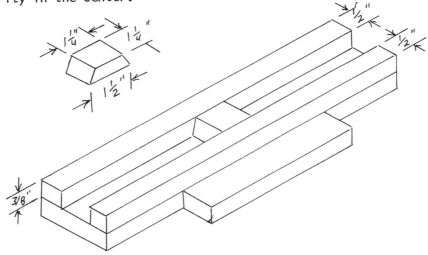

Like the drag half, square it up carefully, and make sure the halves join together neatly.

JOIN THE HALVES

Sand a minimum of draft, both above and below the parting line. Fillet all inside corners, and round all outside corners except at the parting line.

100

Align both halves, clamp them together, and drill the two alignment pin holes through both halves.

Use an 8 penny nail for the drill, and make the pins of the same size nail.

The pins need enter the drag half only enough to ensure proper alignment, and the holes are reamed for an easy fit.

Cement the pins in the cope half, after rounding off the ends so they enter easily.

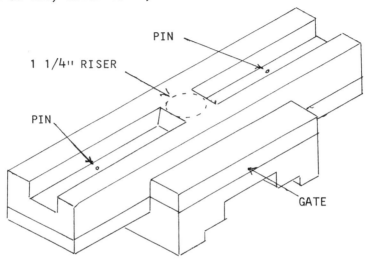

MOLDING THE CROSS SLIDE SUPPORT

Ram up and vent the drag, rub in the bottom board, and roll over. Take extra care to peen the sand into all of the corners and the channel that forms the box slide.

Set the cope half of the pattern, and make sure it will separate easily when you open the mold.

Set a 1 1/4" riser in the center of the pattern, and a 1 1/4" sprue pin about 1 1/2" away from the heavy portion of the pattern.

Ram up and vent the cope, remove the sprue pin and riser, and rub in a bottom board.

Open the mold, and lay the cope down to swab, rap, and remove the pattern.

Swab, rap, and remove the drag half of the pattern, and cut a full size gate to the sprue print.

FIT THE VERTICAL SLIDE

The vertical slide channel is filed and scraped to fit the vertical slide ways in the same manner as the down feed slide. Make diagonal saw cuts to separate the pads, and fit the clamps with shims, as in all box slides.

The clamps are 1/4" X 1" X 3 1/2" cold rolled steel, and each is fastened with three 5/16"-18 X 1" grade 5 cap screws with lock washers.

A 1/8" X 1/4" gib is installed in the left side, and it is adjusted with two 1/4"-20 gib screws with jamb nuts.

The gib screw holes are tricky to drill because the wing of the casting leaves little clearance. You will need to use an extra length 3/16" drill to drill the tap hole. You can get one from U. S. General Supply if you can't find one locally.

The tap holes must be carefully drilled because there is no clearance in the 1/4" slide.

As in all of the gibs, the screws are ground to a point, and they seat in dimples in the gib.

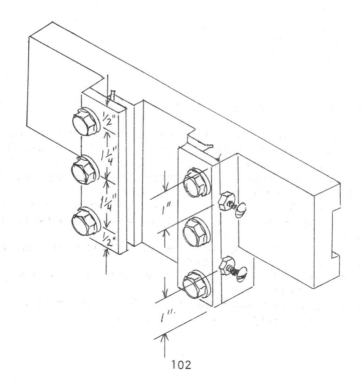

INSTALL THE CROSS SLIDE WAYS

The cross slide support pads are filed and scraped true flat, just like the base of the ram. It does not have to be perfectly smooth, just make sure there is full support under each screw. It may be rough near the center, because of the riser, but don't waste effort perfecting a surface that is not vital.

The cross feed slide ways is a 1/4" X 3" X 1/" slab of cold rolled steel. It is fastened with ten 1/4"-20 X 3/4" flat head screws.

Be certain that the top edge of the ways is 1/2" above the top of the casting, so the cross slide clamp will clear.

The screws are spaced 2" apart, beginning with 1" from each end of the ways. Center the rows in the center of the pads.

Slide the casting onto the vertical ways, and snug up the gib, so you can install the cross slide at exact right angles to the vertical ways.

Install one screw completely, and check the alignment before you install a screw on the opposite end. Check, and re-check.

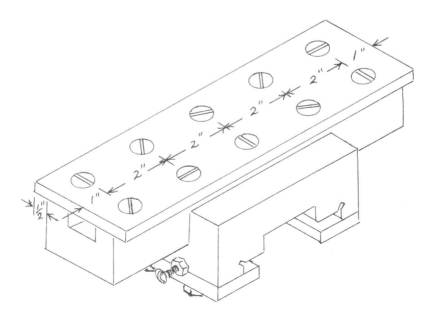

THE CROSS FEED SCREW AND NUT

Two 2 3/4" lengths of 3/8" steel rod, a 7 3/4" length of 3/8"-16 threaded rod, and a standard coupling nut make the feed screw. The coupling nut is cut in half to join the parts together at each end.

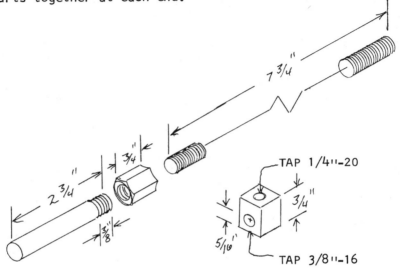

Both ends are the same, and they have 3/8"-16 threads for 3/8" of one end. A cut off bolt won't work as well as the rod, because the shank of standard cap screws is smaller than the thread size.

The feed nut is a 3/4" length of 5/8" key stock with a 3/8"-16 hole tapped through. The 1/4"-20 tapped hole in the top is for the link screw.

Make both ends of the screw alike, and join the parts together with the coupling nuts. The feed nut should be on the screw, and adjust the coupling nuts so there is 8 1/2" between the outside edges of the nuts. Tightening the ends against the threaded rod will jamb the joints to lock the adjustment. You may need to change the adjustment at final assembly, and the coupling nut arrangement makes it easy to adjust the screw for end play as the machine wears in use.

THE CROSS FEED SCREW BEARINGS

A simple shape, but small, and tedious to mold singly. Only two bearings are needed. The third member will be used to make the vertical slide screw.

It's convenient to " gang " such a pattern, and mold all three at once. Just rap the three tails into a blank drag to bed them, and ram up the cope over the common gate. Use a 3/4" sprue pin in the center of the common gate.

Just cut the waste away with a hack saw to leave three separate castings.

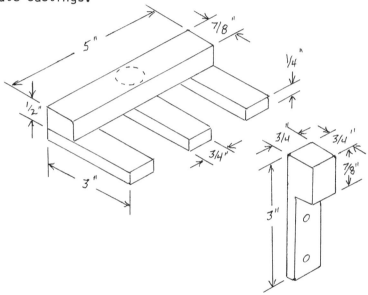

The bearings will be bolted to the cross slide support casting with two 1/4"-20 X 1" cap screws, so drill two 3/16" pilot holes in the tail of each one.

INSTALL THE CROSS FEED SCREW

Clamp a bearing at each end of the support casting, and drill the tap holes in the support through the pilot holes in the bearing tails. Tap the holes in the support for the 1/4"-20 cap screws, and enlarge the holes in the tails to 1/4". Bolt the bearings in place with lock washers on the bolts.

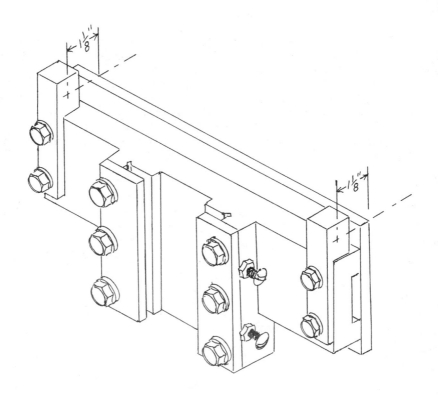

The cross feed screw must run parallel to the cross slide ways, so it is the best reference for laying out the center for the screw bore.

Punch the centers even with the top of the slide, and 1 1/8" from the front surface of the slide.

Drill a 1/8" pilot hole through each bearing, and make every effort to guide the drill parallel to the ways.

Enlarge both holes to 3/16", and slip a 3/16" rod through both holes to check the alignment. Ream if necessary.

Continue to enlarge by steps, until both bores are 3/8" and in line.

Remove one bearing, install the cross feed screw, and re-install the bearing.

THE CROSS SLIDE

A simple casting that needs little discussion.   Being flat at the parting, it's a simple pattern to mold.

The bottom wear pad is wider than the top because of the gib, and the bottom clamp pad is just 1/2" wide for the same reason.

Ram the pattern into the drag, rub in the bottom board, roll over, set a 1" sprue pin 1 1/2" away from each heavy edge, and ram up the cope. Either sprue can serve for the gate, and the other will be the riser.

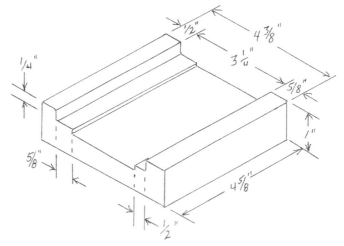

The recess between the wear pads is about 1/16" deep, and the pattern needs minimum draft above the parting line.

The slide is hand fit to the ways, just like the verti-cal slide, and the usual diagonal saw cuts are made to sep-arate the pads.

The clamps are 1/4" X 1" X 4 1/2" cold rolled steel, and they are fit with shims for take up after wear. Grade 5 cap screws fasten the top, and flat head screws hold the bottom.

The gib is 1/4" X 1/8" key stock, and the gib screws are #10-24 machine screws with a point ground on the end. They seat in dimples in the gib.

When the clamps are permanently installed, the entire slide is mounted on the lathe face plate to face off the front surface.

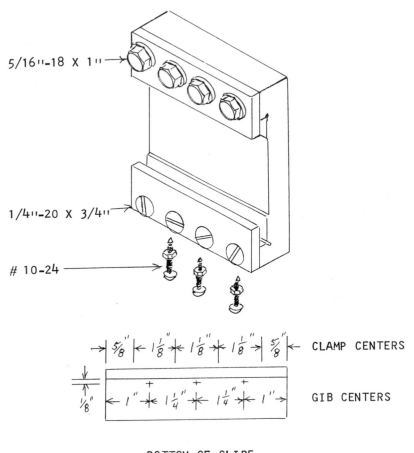

5/16"-18 X 1"

1/4"-20 X 3/4"

10-24

| →|5/8"|← | ←1 1/8"→| ←1 1/8"→| ←1 1/8"→| |5/8"|← | CLAMP CENTERS |

1/8" ←1"→ ← 1 1/4"→ ← 1 1/4"→ ←1"→   GIB CENTERS

BOTTOM OF SLIDE

The hole spacing is the same for both clamps.   Center
the rows in the center of the clamp pads.
    Remove diagonal screws to mount the slide on the face
plate, and use set screw collars for spacers, so the bolt
heads will clear the face plate.   Face off only enough to
clean up the front surface.

THE CROSS FEED DRIVE LINK

A 2" square of 1/16" thick sheet metal with the corner cut away to cl.

It is fastened to the top of the slide with two 1/4"-20 screws. The left hand screw is centered on the slide.

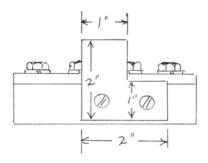

INSTALL THE CROSS SLIDE

Remove the link, to slide the cross slide on the ways, and re-install it to locate the connecting bolt hole.

Locate the center of the cross feed nut on the link, and drill a 1/4" hole in the link.

The connecting bolt is a 1/4"-20 cap screw with a slight shoulder, and all but about four threads cut away.

You can install a 3" ball handle on each end of the screw to try out the cross feed.

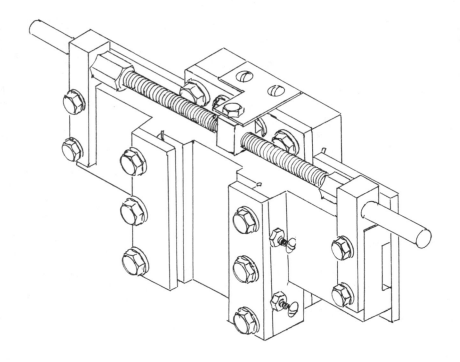

THE VERTICAL SCREW

Made with an 8" length of 3/8"-16 threaded rod, a cap
screw, and another rod coupling nut, like the cross feed.
The extra cross feed bearing casting is altered to serve
as the nut, and special bearings are cast.
The upper bearing is fastened to the column side with
two 1/4"-20 X 1 1/4" cap screws, and the lower bearing will
use 5/16"-18 X 1 1/4" cap screws. A pair of pilot holes are
drilled in the upper bearing, but the hole locations for the
lower bearing are taken from a trial assembly.
The screw is assembled with the bearings and the nut,
and the assembly is clamped in place to align the screw to
the vertical slide ways.
Once aligned, the top holes are drilled through the pilot holes, and the top bolts are installed.

The lower bearing is installed over the existing side bolt, and its center is transferred to the bearing. The forward bearing bolt must be a 5/16", because of the existing hole, but the other bolt can be 1/4" like the top.

The cross slide support casting is drilled through the pilot hole in the nut, and tapped for 1/4"-20 threads. The hole in the nut is enlarged to 1/4", and a 1/4"-20 X 3/4" cap screw is installed.

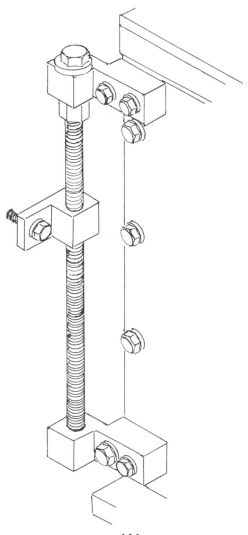

With the bearings loosely clamped to the side of the column, and the nut clamped to the cross slide support, it's an easy matter to tap the members about as you align the screw with the vertical slide ways.

THE VERTICAL SCREW BEARING CASTINGS

Similar to the cross feed screw bearings, and the pattern is made and molded in the same way.

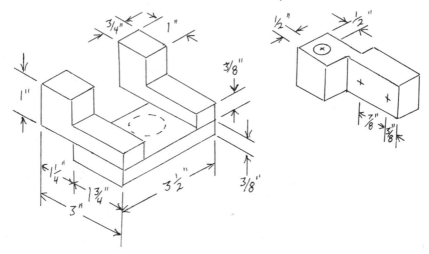

Cut away the common gate, and drill the 3/8" holes in both bearings.

Drill two 1/8" pilot holes in the top bearing, but leave the lower holes to locate during the trial assembly.

THE VERTICAL SCREW NUT

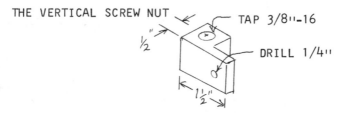

Cut the tail from the extra cross feed screw bearing, and drill and tap the holes.

The cap screw for the upper end should have an unthreaded portion about 3/4" long, and the threads are cut off so the bolt will reach 3/8" through the bearing and two 3/8" S.A.E. washers.

The lower end of the screw passes through the lower bearing merely to guide it.

The fit of the vertical screw is not critical. It must be parallel to the vertical ways, but you can tolerate play in the fit because the vertical screw is not a feed screw. It is used to raise or lower the table to position the job, but it is not moved while work is in progress.

The nut can be turned with a ratchet wrench or a speed wrench to raise or lower the table.

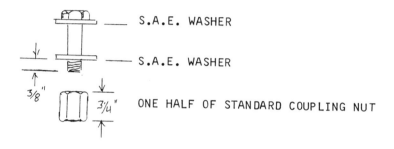

S.A.E. WASHER

S.A.E. WASHER

3/8"

3/4"   ONE HALF OF STANDARD COUPLING NUT

THE WORK TABLE

This is the final casting in the project, and it will be the first item you will machine with the shaper.

It requires a cope that is about 6" deep, and since that is not often necessary, you can make a temporary extension for one of your existing flasks. The drag need only be the usual 3" or 4" deep.

The pattern is of 5/8" section thickness throughout, and it's layed on its open side for molding. A large green sand core forms the cavity in the casting.

When layed on its open side, it looks like a gable roof, and the peak is a hot spot that requires a large riser.

Notice that the panels are tapered on the sides to provide draft both inside and out. This is the same trick we used for the lathe tail stock.

The parting line runs diagonally to the outside corners, so a small portion of the mold cavity will be in the drag. The taper of the panels will take care of the draft on the inside of the pattern, and the outside above the parting line, but you will have to sand a minimum of draft to the ends of the pattern below the parting line.

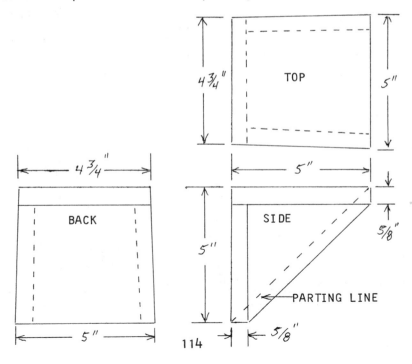

MOLDING THE WORK TABLE

This is a double roll mold, and it begins with a false drag. Prepare a blank drag, and press the base of the pattern in, up to the parting line.

A riser of at least 1 1/4" diameter is cut to fit the peak of the pattern, just like a chimney on a roof.

A 1" sprue pin is set about 1 1/2" away from any side of the pattern, and the cope is rammed up.

Remove the riser and sprue pin, rub in a bottom board, and roll over the entire mold.

Dump out the false drag, slick up the parting face of the cope, and ram up a new drag. Be especially careful to make the green sand core firm, but don't over do it. The core must be strong enough to hold its shape, but it must also be pourous enough to vent, and resilient enough to compress when the casting shrinks.

Vent the new drag generously with the wire, rub in a bottom board, and roll over the entire mold again.

Rap the pattern through the riser hole before you open the mold, so the pattern will remain on the drag. You can slip the riser back in the hole to rap.

Open the mold and set the cope on edge while you rap and lift the pattern from the core.

Clean up, blow out the sprue and riser, cut the runner, and close up the mold to pour.

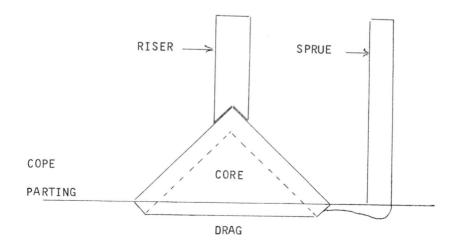

115

INSTALL THE WORK TABLE

The table is fastened to the front of the cross slide with three 3/8"-16 X 1" grade 5 cap screws with flat washers and lock washers.

Both surfaces of the table are prepared the same, so it can be mounted to plane off both surfaces.

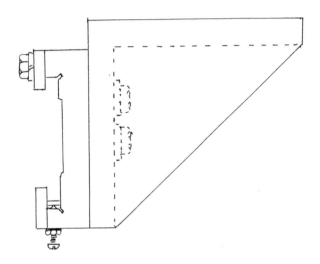

It will be best to prepare a sheet metal template, and use it as a guide to drill 1/8" pilot holes in both surfaces of the table, and also through the cross slide. Then clamp the members together to step drill the holes to tap size.

Do this work with the cross slide removed, so you can clean up the burrs in the slide, and so you don't damage the cross slide ways.

Drill the 1/8" pilot holes through the cross slide and both surfaces of the table, then clamp the table and slide together to step drill the holes to 5/16". Then clamp the other table surface to the slide, and enlarge the second set of holes to 5/16" using the cross slide as a guide.

Enlarge all six holes in the table to 3/8", and tap the holes in the cross slide for 3/8"-16 threads.

Bolt either surface of the table to the cross slide, and re-install the slide. Make sure the bolts don't interfere.

116

SHEET METAL DRILLING TEMPLATE

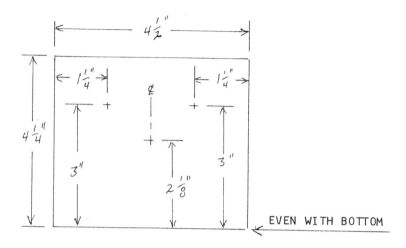

The template is clamped even with the bottom of each of the three surfaces involved, and the base line should be parallel to the top surface of the slide so the table will be planed off with an even section thickness.

At this point, you can operate the shaper by turning the pulley by hand, and actually make a cut on the table.

You can install a couple of ball handles on the cross feed, and operate the cross feed by hand, but it is so tedious that it's best to install the automatic cross feed before you attempt any serious work with the shaper.

CHAPTER VI

THE AUTOMATIC CROSS FEED

This simple ratchet movement is easy to build, and it's action is so fascinating that it becomes the most rewarding part of the project.

The feed cramk is already installed, so part of the job is finished.

HOW IT WORKS

The cross feed screw is a 16 pitch.   That means that one full turn of the screw will advance the cross slide 1/16", or .0625".

The ratchet wheel has 32 teeth, so 1/32 of a turn will advance the cross slide slightly more than .0019", or near .002".

The length of the feed crank stroke can be adjusted to pick up from one to six teeth, so we can have an automatic cross feed of from .002" to .012" on each return stroke of the ram.

The ratchet is re-setting during the forward stroke of the ram, so the table remains stationary during the cut.

The ratchet pawl is reversible, to give automatic feed in both directions.

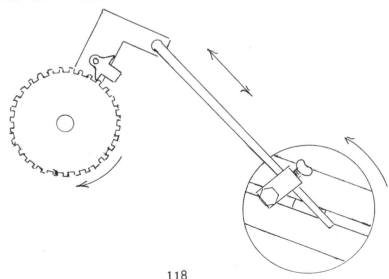

THE RATCHET WHEEL

At first glance, it may seem that this job simply can't be done by simple hand methods, but there is considerable tolerance for error, and the teeth are easily formed with a file.

The cog wheel is made of 1/16" thick steel, and it's fastened to the steel hub with two # 6-32 Machine screws.

The hub has a 3/8" bore to fit the feed screw, and it's locked on with a set screw. You may be able to find an old pulley with a 3/8" bore, and use its hub, or you can drill a 3/8" hole in a scrap of steel and machine the hub on the lathe. In either case, much of the work will be done with the hub mounted on a 3/8" mandrel between centers.

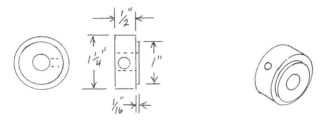

Cut a 3" disc from 1/16" thick steel, and drill a 3/8" hole in the center. Fasten it to the hub with the two #6-32 machine screws. The groove that is formed between the wheel and the hub is for the ratchet plate.

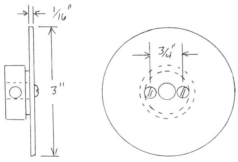

Mount the wheel on a mandrel between the lathe centers, and score a 2 5/8" diameter circle on its face. This will be the pitch diameter of the cog, and it needs to be true.

A row of 32 1/8" holes is drilled at equal intervals around the circumfrence of the circle, and the spacing is just a tiny bit over 1/4" between hole centers.

It will be a help to paint the wheel with layout dye, so the fine marks will show up clearly.

Set a pair of dividers to 1/4", and " step " them around the circle from a marked starting point, to see if you have the right setting for 32 spaces. If the setting isn't right, adjust the divider points and try again. Repeat until you can " walk " 32 steps around the circle, and end on the mark.

Mark the 32 centers carefully, punch them with a sharp center punch, and drill 32 1/8" holes.

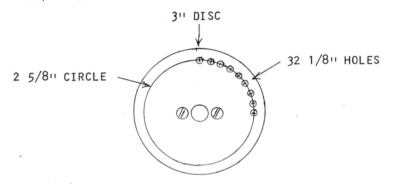

3" DISC

32 1/8" HOLES

2 5/8" CIRCLE

Re-mount the wheel between centers, and cut the disc to the 2 5/8" circle, cutting away half of the holes. Score a sharp line at the base of the holes for a filing guide.

Clamp the disc in the vise, and file each tooth space to the guide line with a 1/8" square file.

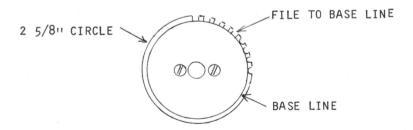

FILE TO BASE LINE

2 5/8" CIRCLE

BASE LINE

The cog wheel need not be perfect, if it appears uniform to the naked eye, that's good enough.

THE RATCHET PLATE

Layed out on 1/16" thick steel, and cut to shape with a hack saw and a file. The outside dimensions are not critical, just make it look good.

The important dimensions are the 1" bore, which fits the shoulder on the hub, and the center location for the pawl pivot pin.

The 1" bore can only be done accurately on the lathe, and you can do it by mounting the plate on a scrap of plywood with two #10 X 3/4" sheet metal screws, and clamping it in the face plate angle clamps.

Center it with the tail stock center, drill a starting hole with the tail stock chuck, and bore to fit the shoulder of the hub with the boring bar.

You can bore tight through the plywood, so you can get your calipers or gauge in the hole as you progress.

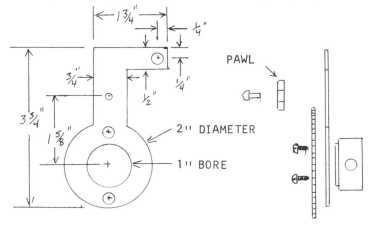

The pawl pivot hole is 1/8", the upper link hole is 3/16", and the two holes near the bore are drilled to suit the screw you use to mount the blank on the plywood.

When the plate is installed on the hub, it should swing freely, with barely perceptible play. You may have to cut the shoulder a bit deeper, or face off the end of the hub to bring it to a proper fit.

THE RATCHET PAWL

Made from a scrap of 1/8" X 3/4" steel, 1 1/4" long. It is fastened to the plate with a 1/8" tubular rivet. It is meant to swing freely, with barely perceptible play.

1/8" HOLE ⟶

THE CONNECTING LINK

A 10" length of 3/16" steel or brass rod, with one end threaded, and bent at right angles.

|←——————————— 9 1/2" ———————————→|

1/2"

10-24 THREADS

THE CRANK BLOCK

A scrap of aluminum or a piece of 5/8" key stock. Bore the 3/8" hole in the lathe, but the others can be drilled.

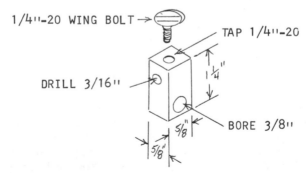

1/4"-20 WING BOLT ⟶

TAP 1/4"-20

DRILL 3/16"

1 1/4"

5/8"

BORE 3/8"

5/8"

THE FEED CRANK PIN

A 1/4" X 3/8" brass or steel bushing, two 1/4" S.A.E. washers, and a 1/4"-20 cap screw.

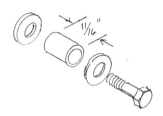

THE FINAL STEPS

Install the ratchet assembly on the cross feed screw, and install the 3" ball handle about 1/2" away from the ratchet. Cut off excess shaft length, and file a flat on the shaft for the set screws to seat against.

Install the connecting link with two # 10-24 nuts tightened together as jamb nuts.

Slip the crank block on the connecting link, and bolt it to the crank plate with the cap screw and sliding nut.

It may be necessary to bend an off-set in the connecting link to bring it parallel to the crank plate.

The ratchet is meant to turn the feed screw on the return stroke of the ram. You can turn the pulley by hand to determine the correct position to lock the crank on the shaft. position it so there is no motion to the screw during the forward stroke of the ram.

CHAPTER VII

THE COUNTERSHAFT

The design and construction is identical to that used on the metal lathe in book 2 of the series. Only the pulley sizes are changed because of the different speeds used.

A 1 1/2" pulley is used on a 1725 RPM motor, and an 8" pulley is used on the countershaft.

The cone pulleys are 4", 3 3/8", 2 5/8", 2". The ram strokes per minute are approximately 40, 62, 103, and 161.

The lathe countershaft can be used just as it is, but only the two lowest speeds will be of use. The higher speeds are so fast that the clapper won't re-seat on the forward stroke.

The shaper needs to be very securely bolted to a sturdy bench top. The momentum of the ram exerts considerable force at the end of the stroke at higher speeds.

The next several pages are mostly a repeat of chapter VI in " METAL LATHE ".

Being of simple design, little needs to be said about the construction of the frame assembly. Of course you can use different sized angle and strap iron. The sizes in the plan were what I had on my junk pile.

Welding would be the fastest way to join the members, but if you don't have a welder, use iron rivets instead of bolts and nuts. They are easy to cut with a hack saw, the ends peen easily, they are cheaper than bolts, and they don't work loose.

THE OVER CENTER LOCK

The mechanics of the assembly are simple. Notice the dotted center line from the release link pivot to the release lever pivot. When the release lever is drawn forward, the crank will pull the lower link pivot over the center line, unlocking the linkage, and pulling the upright forward. When the handle is pushed back, the end of the crank will be against the link pivot bracket, stopping the motion, and locking the action in the over center position. A single forward motion releases belt tension for a speed change.

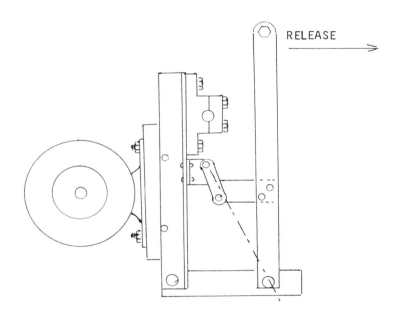

RELEASE

To determine the right size belt, just run a flexible tape measure over the outside of both pulleys, as though it were a belt, and this gives you the length. Use " A " size belts, which are 1/2" wide.

The design is adaptable to most fractional horse power motors, but the spacing of the motor mounting bolts will be different. The dimensions given are for a NEMA 48 frame, and yours may be different, so check.

It will probably be best to build the motor mount to fit your motor first, since you will need to know the dimensions to make the spreaders for the frame.

The motor mounting rails are two pieces of 1 1/4" X 1/8" angle iron, 6 1/2" long. They are drilled and tapped for 1/4"-20 threads on the spacing that corresponds to the vertical spacing of the motor base slots.

The motor rail spreaders are 1" X 1/8" steel strap, cut 1" longer than the space between the slot centers on the motor base. They are drilled 1/4", on centers 1/2" from the ends of the spreaders.

By bolting the spreaders to the rails, and letting the bolts extend through the rails about 3/4", you will have studs for mounting the motor.

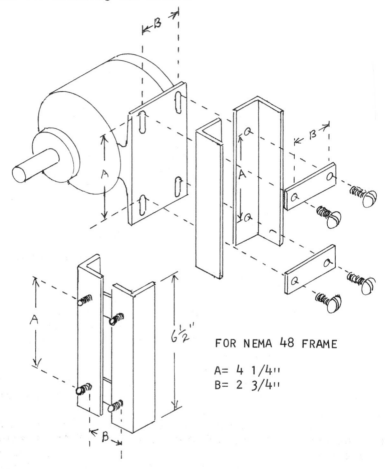

FOR NEMA 48 FRAME

A= 4 1/4"
B= 2 3/4"

The base spreader is of the same material and dimension as the motor rail spreaders, but it is riveted to the base rails through countersunk holes in the bottom of the rail.
The spreaders are the last order of construction in the frame assembly.

To clarify the exploded drawing, the parts are named as follows:

1. Motor rail, 1 1/4" X 1/8" angle, 6 1/2" long.

2. Motor rail spreader, 1" X 1/8" strap, 3 3/4" long.

3. Base rail spreader, 1" X 1/8" strap, 3 3/4" long

4. Upright rail, 1 1/4" X 1/8" angle, 10" long.

5. Base rail, 1 1/4" X 1/8" angle, 6 1/2" long.

6. Release crank, 1" X 1/8" strap, 3" long.

7. Release lever, 1" X 1/8" strap, 13" long.

8. Release link pivot, 1 1/4" X 1/8' angle, 1 1/4" long.

9. Release link, 1/2" X 1/8" strap, 2" long.

10. Release handle, 1" O. D. pipe, 3 3/4" long.

11. 8" outboard pulley.

12. Pillow block bearings.

13. 4", 3 3/8", 2 5/8", 2" step cone pulley.

14. Set screw collars.

It will be easier to build the unit in two parts, as shown in the exploded view.   Having determined the proper bolt spacing for the motor rails, set the spreaders aside for now, and make the right and left hand sections of the frame.
The pivots for the uprights and release levers should be 1/4" rivets. The link pivots, and all others are 3/16".
Make the base rails first.   Drill 1/4" holes in each end, for the mounting slots, and cut away the waste with a hack saw.   Drill the pivot holes and the spreader hole, counter sink the spreader hole on the bottom of the rail, and set the rails aside.
Assemble the uprights and motor rails, drill and tap for the pillow blocks at the top of the upright, drill the pivot

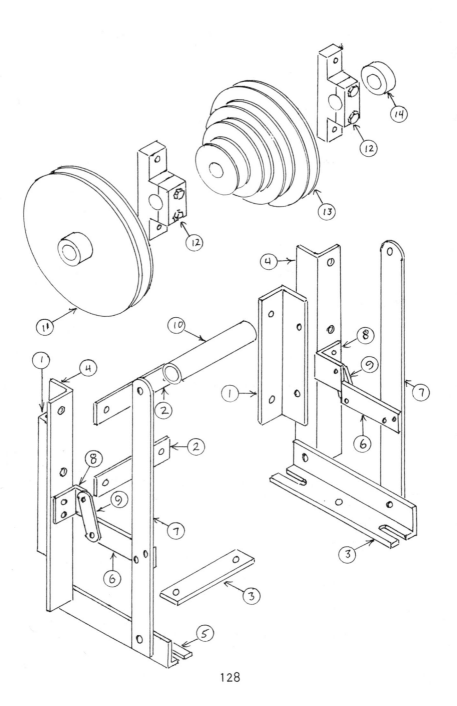

hole, but don't drill the rivet holes for the release link pivot. Rivet the upright pivot to the base rail.

Assemble the release lever, the release crank, the release link, and the release link pivot. Rivet the release lever pivot to the base rail.

Clamp the release link pivot to the upright, drill the rivet holes through both members, and set the rivets.

Keep in mind that there is a right and left hand member, you are making opposing sides.

Install the motor rail spreaders with the bolts extending through the tapped holes in the motor rails.

Install the base rail spreader with rivets beat flat in the counter sunk holes in the bottom of the base rails.

Install the release handle with a length of 1/4" threaded rod. Fit a wooden bushing in each end of the pipe to center the bolt.

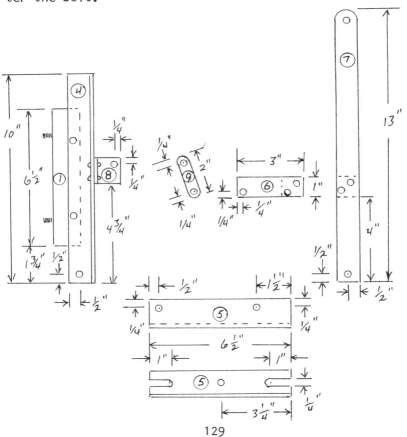

129

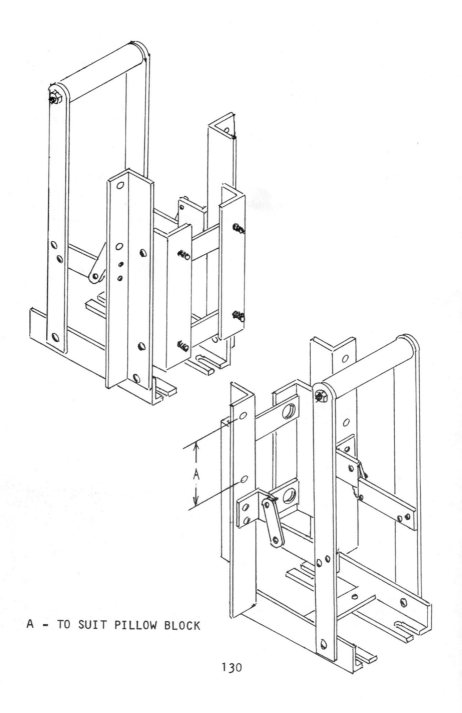

A - TO SUIT PILLOW BLOCK

130

FINAL INSTALLATION

With the machine firmly bolted to a sturdy bench top,
align the counter shaft so you are sure the ram will clear
on the back stroke. Position the bolts so you will have an
amount of adjustment when the belt streches and wears.

SAFETY FIRST

Open belts, chains and sprockets, and reciprocating arms
are extremely dangerous. It would be very wise to devise
a set of guards to protect yourself and others from injury.
The reciprocating actions in the machine are extremely
powerful because of the high ratio of reduction. They seem
safe and innocent because they move so slowly, but they are
capable of shearing off a finger, or even an entire hand.
Install a switch at a point that can be reached quickly
in case there is any problem during operation.

CHAPTER VIII

CONCLUSION

PLANE THE WORK TABLE

Both surfaces of the table should be planed off, and it doesn't make much difference which is first.

The machine was designed so the ram will retract into the column to enable it to plane its own table. The down slide must be raised above the ram slide for this operation, and the rotating head must be in the straight up position. It is probably a good idea to re-position the ram clamp guide after you have planed the table, so you won't forget that the travel is limited.

A standard 5/16" lathe tool bit is 2 1/2" long, and it will reach the table with the down slide raised high enough to clear the ram slide clamps. You can widen the slot in the tool post for a 3/8" bit, which is 3" long, if you need additional reach for any purpose. You may even want to make an extra tool post for the larger size.

The normal travel of the table is from left to right, but it will work as well from right to left.

Be sure to operate the shaper through one or more full strokes by hand before you turn on the power, to be sure it is properly set up. This should always be the last step before you turn on the power at any time.

GRINDING TOOL BITS

This may seem to be a complicated art, and in modern high production operations it is a science. There is little point in talking about precise angles when there is no means to gauge those angles, so we'll just talk about free hand tool grinding.

Having decided on the shape of the cutting edge, whether flat, pointed, or round, you need only be concerned with the back rake and end relief.

Back rake should be from 3 to 5 degrees, and end relief is from 5 to 10 degrees.

These angles are judged on the vertical plane, as the bit is installed in the tool post.

Side rake and side relief are of importance on finishing tools, but the angles are small, and require little more than a touch on the grinding wheel.

Side rake would normally be no more than 5 degrees, and side relief would be about 3 degrees.

Shaper tools are ground much like lathe tools, but the rake and relief angles are less, so they won't dig into the work and distort the machined surface.

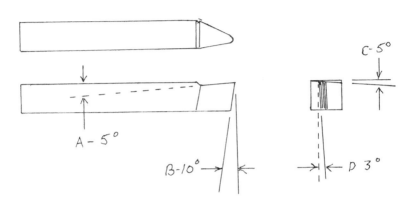

A= BACK RAKE
B= END RELIEF
C= SIDE RAKE
D= SIDE RELIEF

The illustrated tool is a good general purpose tool for roughing and finishing a flat planed surface. The form is a slightly rounded nose, and all of the rake and relief angles are extreme.

Special forms may require more side relief, especially those ground for cutting keyways, splines, and gear teeth.

Excessive back rake and end relief cause the tool to dig in, and excessive side rake can cause the tool to pull off the cutting line.

Tools for cutting slots, such as keyways and splines, are slightly wider on the cutting edge, and they have relief on both sides.

Special forms, such as gear teeth, may require a roughing cut with a slotting cutter before the final shape is cut.

In general, the rake and relief angles are reduced as the cutting edge increases in size. Pointed tools can have more acute angles than blunt nosed tools.

Experimentation is the best teacher. The job at hand will test your judgement, and you'll quickly learn how to grind the shape you need.

A few strokes with a whetstone will make the cutting edge last longer and cut smoother.

ADJUSTING THE STROKE LENGTH

Clamp the work as near to the column as possible, and begin the stroke about 1/2" from the work. The tool should travel about 1/4" beyond the work.

Always rotate the shaper through one full forward and return stroke before you turn on the power.

CLAMPING THE WORK

Many jobs can be bolted to the table, just like you do on the lathe face plate. Bolts, clamps, and furniture can be used.

Now that you have both a lathe and a shaper, you can make just about any kind of jig or fixture that you might need to hold unusual shapes.

You can cast a blank and make a Vee block for holding round work for machining flats and keyways.

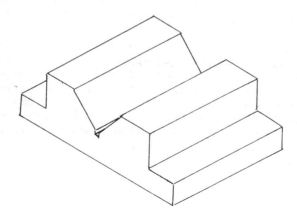

You can cast some 1 1/4" square stock and make a pair of clamp dogs to bolt to the work table as a vise.

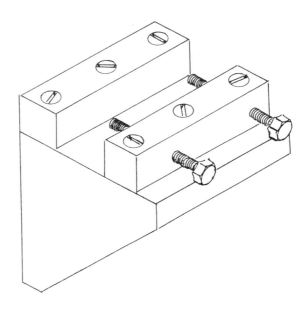

Of course they can be mounted on the table at any angle, and you can use the shaper to plane them true square. If you face off two sides parallel in the lathe, the shaper can finish the inside clamping surfaces true to the table.

It is well within your means to make a very good vise on a swivel base, so you can mount work at any angle without drilling extra holes in the table.

Many of the special holding jigs and fixtures that you make for the shaper table will work equally as well on the lathe face plate, so each time you solve the problem at hand, you also solve future problems.

You can make your own reamers, and flute them with the shaper, and when you add thread cutting change gears to the lathe, you can make special taps, such as for left hand threads, or for acme or square threads.

135

You can mount fluting jobs in Vee blocks, and estimate the divisions for a one only job, or you can mount them between centers, and use an indes plate for dividing.

We'll get into the construction of an accurate dividing head in book 6, but for now you can make a simple indes plate in the same way you made the cog wheel for the ratchet. The divisions don't need to be absolutely perfect for home made reamers, and you can lay out a 12 hole index plate with the simple hand method that will be accurate enough for many of these simple jobs. With it, you can have 2, 3, 4, 6, or 12 divisions, and that covers many needs.

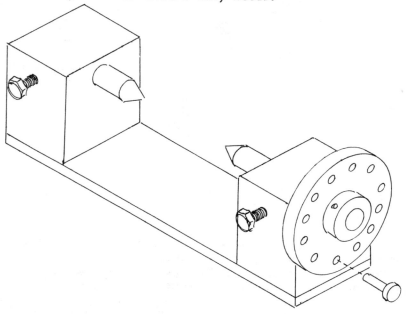

The illustration shows a dividing fixture in elementary principle. The head stock and tail stock can be simple in shape, and fastened to a slab of 1/4" X 2" cold rolled steel with flat head machine screws. Both blocks can be bored alike, and aligned with a close fitting shaft when they are fastened to the bed.

The side clamp bolt on the tail stock can be used to set the center distance, and the one on the head stock will lock the index spindle on its setting to make a cut.

The center height can be about 2", and the spindles are 5/8" diameter with accurate 60 degree cone centers. The tail center needs a flat for the set screw, and the index center needs a groove, like the ram head arbor, so the set screw won't jamb up the bore. The index center arbor should be machined with a shoulder for thrust, because a set screw collar would be likely to slip under the strain of cutting.

The index plate can be cast in aluminum, and machined on the face plate of the lathe. If you scribe a sharp clear line for the hole circle, and step drill 3/16" holes carefully, the plate can be extremely accurate.

The loose pin passes through the plate hole and into the head stock casting. You can make an oversize pin to fit the drilled holes closely. The pin only registers the index on the division, the side set screw locks the arbor for the cut.

The work is mounted between the centers, and a slotted drive plate on the index arbor engages a dog on the work.

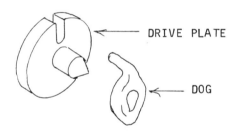

DRIVE PLATE

DOG

Some work can be mounted on the index arbor with a set screw. You'll find plenty of work for this simple fixture, and you'll soon have a set of arbors for mounting the work.

You can elaborate on the design, and adapt it to special needs. The plate can be screwed to the head stock, and the pin carried in a quadrant, just like the commercial models.

Commercially made fixtured have spindles with taper shank centers, and they're supplied with a set of precision plates. Most that I've seen are too large for the home shop, and they are certainly priced too high for my purse. Furthermore, our work rarely calls for such precision as is built into them, and if it does, we can better afford the time than the money.

A VERSATILE MACHINE

A metal shaper is capable of all manner of straight line machine work. It can actually do nearly anything a milling machine can do.

Because it does its work with a single point tool, one stroke at a time, it is not as fast as a miller, and so it has fallen in disfavor with modern industry.

In addition to straight line work, It can also cut concave and convex shapes, and irregular profiles. It's beyond the scope of this manual to describe all of the accessories that can be applied to the machine, but you can learn much from machinists hand books that were written a generation ago.

English history claims James Nasmyth, a pupil of Henry Maudslay, as the inventor of the shaper. Its original purpose was to form the slides for machine tools in a more rapid and accurate manner than the hand methods that we began with in these projects. We're free of that burden now, We've made the same stride that was made by Nasmyth in about 1850, and it doesn't matter that we're about 130 years late.

The basic principles in the design of the machine have remained unchanged, and other machines like the miller and precision grinders have evolved on the principle.

As you become familiar with the machine, and assign it to various special jobs in your shop, you will be impressed with its versatility, and you'll understand why it's the logical third step in building your own machinery.

A METHOD TO THE MADNESS

By now you have seen that the series is systematically exposing the simplicity of machine tools and metal working techniques.

It has become a tradition to cloak these things in deep mystery. We're supposed to believe that the knowledge and skill that is required is so elusive that only a few special people can acquire it, and the rest of us must pay others to do this highly specialized work. That may be fine for the specialist, but some of us can't pay the price.

Having began with a simple home foundry, and developed your wood carving skill into metal casting, you've built a metal lathe, and used it to machine the parts for the metal shaper. Most likely, these were machines you would not have bought, because they simply cost too much, and you have no real need for them, you just want them.

The next step in the project is the milling machine, and the metal shaper will be a great help in building it. Not only the machine, but the skill and knowledge you have gained in building it.

Some of the castings in the miller will be more complex, but you will be able to handle them because of the foundation you have laid so far.

Likewise, some of the mechanisms in the miller would have been more difficult to understand, and the parts would have been tough to form without the help of the shaper.

The miller will broaden your shop capacity to handle a wide range of jobs that you may not have dreamed possible. Large diameter turning and boring are jobs that a miller does very well. There are also some large facing jobs that may be beyond the capacity of the shaper. These can be done on the miller with ease.

Both the shaper and the miller will be used to add the back gears and change gears to the lathe, and both machines will find plenty of work in building the drill press.

Such accessories as the dividing head and a four jaw chuck for the lathe would be near impossible without one or the other.

Each phase brings new problems and new rewards, and we grow as we go.

As a matter of passing interest; I've been able to find just one company that offers a metal shaper. It's an 8 inch bench model, and the price is $3,600.00. The brand isn't mentioned, so I don't know if it is American made or imported. I would not have been able to buy it, even if I'd known it was available, but seeing it in the catalog makes me feel I was well paid for building my own.

SUPPLY PROBLEMS

The great search goes on for a scource of supply for the incidental items that are needed in the shop. I had hoped to be able to include more supply houses that would answer mail orders for small users, but not much luck so far.

Sears tool catalog lists a number of items that are not in their general catalog, and I've relied heavily on them for some of my shop items.

U. S. General Supply Corp., 100 General Place, Jericho, N. Y. 11753, has a large line of tools. At the last writing, their catalog was $1.00. I've ordered a number of items from them, including my vernier calipers, and have been well satisfied.

If you must buy some of your cold rolled steel new, shop around, because there is a wide range of prices on steel. It is not unusual for a shop to charge from $15.00 to $25.00 just to cut your steel from a full bar. You can buy a full bar for less than the cutting charge in many cases. I paid about $35.00 for all the steel I needed for these projects, and I have some left over for future work.

W. W. Grainger is a wholesale house, and they will not sell to individuals at retail, but you can usually buy in the name of your employer. The purchasing agent in your company is likely to have a catalog.

Brodhead Garret is a large outlet that caters to schools and public institutions, but they also sell to individuals. Their minimum order amount is $25.00. They accept Master charge and Visa Credit cards, and they have Toll-Free order numbers. In the west, it's 800-824-7900, in the North East, it's 800-321-6730, In the South East, it's 800-841-4967. Their main office is 4560 E. 71st St., Cleveland, Ohio 44105. They have a number of foundry specialty items, and near any thing you can imagine in a catalog of over 1000 pages.

J. S. Whiteside, P. O. Box 730, Pineville, N. C. 28134, will sell foundry supplies in small quantities to individuals.

Other scources have not yet responded to inquiries, so we'll have to wait and see what happens.

Local industrial supply houses will usually have a better line of drills, taps, and cutting tools than retail hardware stores, and the price may not be much higher. Good tools are worth the difference in most cases.

If any one knows of companies that will meet the needs of the home shop, I'd like to have the information to pass along.

In the mean time, devise substitutes, and if you have a problem, drop me a line and I'll try to help. There is certain to be an answer, and it will be simple.

INDEX

BUILD YOUR OWN METAL WORKING SHOP FROM SCRAP

A progressive series of projects,   planned around home made castings in aluminum and pot metal.

Beginning with ordinary hand tools, and improvised equipment, the scrap metal is transformed into practical machinery for your home work shop.

The sequence is planned to show you the latent ability in untrained hands, and develop it into valuable skill.

The designs are simple, and the procedures are within the ability of a novice.

Special treatment is given to pattern making, molding, and casting the parts. It really is simple.

No outside custom machine work is required. The work is done with the machines you build.